THE RAVEN'S HAT

THE RAVEN'S HAT

Fallen Pictures, Rising Sequences, and Other Mathematical Games

Jonas Peters and Nicolai Meinshausen
Illustrated by Malte Meinshausen

The MIT Press
Cambridge, Massachusetts
London, England

This book was set in Stone Serif by Westchester Publishing Services. Printed and bound in the United States of America.

Library of Congress Cataloging-in-Publication Data

Names: Peters, Jonas, 1984– author. | Meinshausen, Nicolai, author. | Meinshausen, Malte, illustrator.
Title: The raven's hat : fallen pictures, rising sequences, and other mathematical games / Jonas Peters and Nicolai Meinshausen ; illustrated by Malte Meinshausen.
Description: [Cambridge, Massachusetts] : [The MIT Press], [2020] | Includes bibliographical references and index.
Identifiers: LCCN 2020004662 | ISBN 9780262044516 (paperback)
Subjects: LCSH: Mathematical recreations. | Games in mathematics education.
Classification: LCC QA95 .P4335 2020 | DDC 793.74—dc23
LC record available at https://lccn.loc.gov/2020004662

10 9 8 7 6 5 4 3 2 1

To my dear Wangerooge crew.
—Jonas

To my dear Zurich-Freiburg-Melbourne family.
—Nicolai

CONTENTS

PREFACE AND ACKNOWLEDGMENTS

Der Mensch spielt nur, wo er in voller Bedeutung des Wortes Mensch ist, und er ist nur da ganz Mensch, wo er spielt.[1]

—Friedrich Schiller

This book contains games that are special in the following sense. When confronted with them for the first time, most people (including us) consider them unsolvable. Surprisingly, this perception changes when the games are translated into mathematics. The mathematical language nicely reveals the problem's underlying structure. Existing mathematical theory then allows the reader to quickly solve the game.

For us, understanding how mathematics unfolds the game's solution were moments of joy, and we wrote this book in the hope that the reader will experience the same. We therefore invite any reader to join us in being astonished about the fact that the game is solvable, in thinking about possible solutions, in having ideas that unfortunately do not work, and in enjoying the moment of understanding.

During the presentation of solutions, we mainly focus on conveying mathematical concepts and ideas. In particular, we explain the mathematical concepts using many examples. We aimed at maintaining mathematical precision, but we also avoided overly complex mathematical notation. This way we hope that

1. The quote translates roughly as "Humans play only where they can be truly human, and they are fully human only when they play."

readers, ranging from high school students to those dealing with mathematics in their professional lives, can find pleasure when reading the book. Readers who are used to concise mathematical language can find additional details in the appendixes and pointers to the relevant literature.

All of the games can be performed in front of an audience. We have added some sections with practical advice, but interested readers can certainly complement these suggestions with their own ideas on how best to perform the games. The games can be presented in about 60–90 minutes, which makes them suitable for seminar talks or special lectures, such as Christmas lectures.

Most of the games described and their solutions are well known in some form or another. In places, we took the liberty of slightly changing the setup or the presentation of solutions. We did our best to supply information on their origins, as far as this is known to us. We refrain, however, from providing extensive references in the main text (e.g., regarding mathematical concepts and results) so as to not disturb the flow of the text more than necessary.

Each chapter is self-contained in that it contains a complete description of the problem, an introduction to the relevant mathematical theory, some notes on the history of the problem (to the extent that we are aware of it), and in some cases, comments on variations of the game and practical advice for a group performance. It is therefore possible to start reading with any chapter.

The majority of the games include some kind of randomness, and we have implemented some of the games in the programming language R. The code can be downloaded at https://github.com/mathemagicalgames/ and can be used for simulating games and their winning strategies.

Mathematics allows us to express thoughts and arguments in a concise and unambiguous way. Some math expressions are used so frequently that it is worthwhile to introduce notation for them. The ones that we use throughout this book are collected in appendix A. The list there can be checked when encountering some unknown expressions—we hope that we have explained all relevant terms in that appendix. The experienced reader will realize that the list mainly contains standard notation. Appendix B contains further information on binary numbers, the convergence of sequences and series, and the exponential and logarithmic functions, for example. These concepts are widely used in mathematics, and readers who have not been introduced to them before might benefit from reading these chapters. Appendix C contains further details on the individual chapters. We expect this information to be most valuable for experienced readers who have been exposed to university level mathematics.

We hope this book appeals to people who are already interested in mathematics but also to people who are mainly interested in performing the games (and who might still take a look at the maths during some dark winter evening).

We are very grateful to several people without whom this book would not exist. Jonas thanks Wolfgang Merkle for his seminar at the University of Heidelberg in 2006, Roland Langrock and all participants of the Deutsche Schülerakademie 2009 and 2010 for two amazing summers in Rostock, and Anders Tolver for a lot of fun at the Christmas lectures at the University of Copenhagen. We thank ETH Zurich and the University of Copenhagen for supporting this project; Bryony Leighton for proofreading; Anders Tolver for many helpful comments; and Malte Meinshausen, who created the beautiful drawings for this book.

Copenhagen and Zurich, December 2019

1 HAT COLORS AND HAMMING CODES

1.1 THE GAME

Number of players:	3, 7, or 15
You will need:	2 colored hats or caps for each player (1 in blue and 1 in red); alternatively, colored tags that can be attached to the hats

The players sit in a room, wearing red and blue hats. Each player's hat color has been chosen by members of the audience. Every player is able to see all the hats except their own. Below is an example of Anwar, Bella, and Charlie sitting in a circle, with Anwar wearing a red hat and Bella and Charlie wearing blue ones. The players are not allowed to talk to one another, and after a few seconds, they have to hold up a sign with their answers to the question: "What is the color of your hat?" The answers are "red," "blue," or "?" (which corresponds to "I don't know"). All players win or lose together. They win the game if and only if at least one of the answers is correct and no player gives a wrong answer. Question marks never count as incorrect, but wearing a blue hat and answering "red" is wrong, and so is answering "blue" when wearing a red hat. If, for example,

Bella and Charlie correctly answer "blue" and Anwar says "I don't know," they have won the game. But if all 3 players answer "blue," they lose (because Anwar's answer is incorrect).

What is the best strategy? How likely is it that the group of three players will win the game? What if there were n *players?*

Let us first assume that the audience distributes the hats at random, that is, the color of one player's hat does not contain any information about the color of any of the other player's hat, and each color is chosen with the same probability. It is perhaps surprising that there is an interesting strategy at all!

The group of players could, of course, decide on one person (e.g., Anwar) to always guess "blue." If the others answer "I don't know," the probability of winning as a group is 1/2. Anwar can also choose "red" or "blue" at random, but his chance of getting the color correct would still be just 1/2, so the group will still only win half of all games on average. The group could decide to let more of the players guess their color, either at random or by picking a color beforehand. The chance of winning in this way would be just 1/4 if two players are allowed guess the color. Both players announcing a color would need to be correct, and each of them is correct with probability 1/2. The chance that they are both correct is then the product of the individual probabilities, which is 1/4. This drops to just 1/8 if three players are allowed to pick their colors. Letting just one player pick his or her color at random is better than letting more people pick their colors at random.

But what else could Anwar do instead of just picking the color at random or in a predefined way? Say that he bases his decision on the colors of Bella's and Charlie's hats. These colors do not contain any information about the color of Anwar's hat, as his color was chosen independently of the other colors. In other words, regardless of any observed sequence of the colors of Bella's and Charlie's hats, the color of Anwar's hat will still be blue or red with equal probability.

We encourage the reader to pause here and think again about the questions presented on the previous page.

1.2 HOW WELL CAN A STRATEGY WORK?

Individually, we have no predictive power, and any guess of an individual's hat color is equally likely to be correct or wrong. Therefore, the key to success will be to "collect" the wrong answers in single instances of the game. That is, the players need to ensure that either just one person is correct (and the others say they do not know) or that all players announce their colors falsely, thus effectively bundling the false guesses into the same game and spreading out the correct guesses over as many games as possible.

We are now going to introduce some notation that will later help us to understand the general solution. We first associate

blue with a 0 and red with a 1.

In the example, the colors (red, blue, blue) for players 1, 2, and 3 (Anwar, Bella, and Charlie, respectively) are then equivalent to the vector $(1,0,0)$. In the following, we will simply write (100) and call this a sequence rather than a vector. This notation works, too, of course, if there are n players in total. The players' hat colors can then be described by a sequence with n entries. These sequences look like (0110100011), for example, if $n=10$.

If the players choose a deterministic strategy, their answers will always be identical if they see a specific sequence of colors on the other players. For each true sequence x of colors, we then get one collection of answers for Anwar, Bella, and Charlie. These answers can be represented in a matrix.[1] Its rows correspond to the 8 different color sequences x (each occurring with probability 1/8), and each column shows the answers given by one of the three players. Any strategy the players adopt will lead to outcomes that can be represented as follows:

true sequence		answers of Anwar	answers of Bella	answers of Charlie		group win or loss
$x=(000)$	→	correct	no answer	correct	→	win
$x=(001)$		no answer	correct	wrong		loss
$x=(010)$		no answer	no answer	wrong		loss
$x=(100)$		wrong	wrong	correct		loss
⋮			⋮			⋮

Here, only the first row corresponds to a successful run, as at least one player has answered correctly and nobody gave a wrong answer. In the example, Anwar has decided to guess his color only if the colors of Bella's and Charlie's hats agree. He then goes for the same color. In contrast, Bella copies Anwar's color but only if Charlie does not also have the same color. And Charlie always copies Bella's color.

In the first game—that is, the game represented by the first row—both Bella and Charlie have a blue hat on, and Anwar

1. A matrix is a table with rows and columns. See appendix B.7: "What Is … a Matrix?"

guesses "blue," which happens to be right. In the fourth game, Bella and Charlie also have blue hats on, but Anwar's guess turns out to be wrong. He decides not to provide an answer in the second and third games, as Bella's and Charlie's colors do not agree. From Anwar's point of view, the first and fourth game settings look identical. If he decides to guess a color (rather than saying "I don't know"), one of these game settings will yield a correct answer and the other row will yield a wrong answer. Thus for each player (i.e., for each column of the matrix above), the number of right answers equals the number of wrong answers, which again shows that, working as individuals, the players have no predictive power. If adopting a randomized strategy, each individual player will be no more likely to make a correct guess than a wrong one.

If this constraint (i.e., that there are the same number of right and wrong answers in each column) were the only constraint, how well could we do? The key is to bundle all the wrong answers into the same row and spread out the correct answers among as many rows as possible. A good arrangement would, for example, be the following:

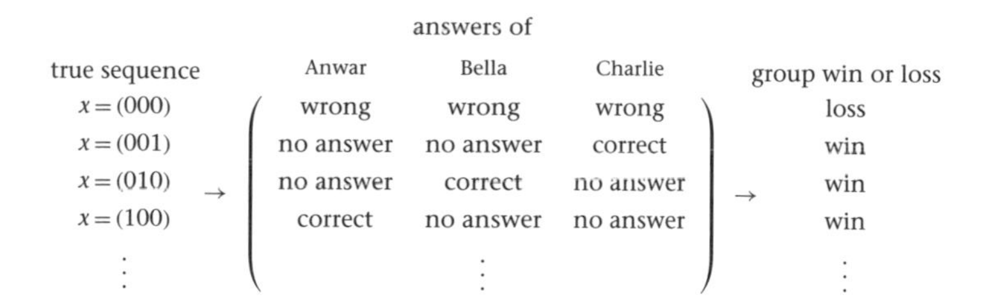

true sequence		answers of Anwar	Bella	Charlie		group win or loss
$x = (000)$		wrong	wrong	wrong		loss
$x = (001)$		no answer	no answer	correct		win
$x = (010)$	$\rightarrow$	no answer	correct	no answer	$\rightarrow$	win
$x = (100)$		correct	no answer	no answer		win
$\vdots$			$\vdots$			$\vdots$

Here, all players guess their hat color incorrectly in case of the first color sequence (000) but then spread out their correct answers among the other color sequences (along with "I don't know" answers from other players), thereby maximizing the number of games they win. For any row with a wrong answer, we get, at best, three rows with a correct answer. Therefore, the group has to lose in at least one in four equally likely color sequences (since rows without any answer are lost, too). So for any strategy,

$$P(\text{group loses}) \geq \frac{1}{4}.$$

With n instead of three players, the same argument yields

$$P(\text{group loses}) \geq \frac{1}{n+1}.$$

To obtain this bound, we have only used the fact that the hat color of a given player is independent of all the other hat colors, which led to the constraint of equal numbers of right and wrong answers for each player. Can we find a strategy for which the probability of losing comes close to this bound?

1.3 SOME MATHEMATICS: HAMMING CODES

Let us suppose that we would like to send a message from a sender to a receiver over a noisy channel, as if we were making a call on our cell phone, which is being disrupted by atmospheric turbulence and flocks of birds. Suppose we only want to send a sequence of letters $\{a, b\}$ (e.g., the sequence *abba*). In digital systems, we translate this message into a binary sequence containing only 1s and 0s. This step is called "encoding," and it works for many practical applications: We can then send the message using a digital system with 0 and 1 being low and high voltage levels, respectively, for example. However, most physical systems are noisy, and we can now try to find a binary code that will enable us to detect the noise in the transmission and remove it, at least partially.

Suppose we encode a as (000) and b as (111). Then the transmission of a sequence *abba* would look like

$$abba \overset{\text{encoding}}{\longrightarrow} 000\ 111\ 111\ 000 \overset{\text{noisy channel}}{\longrightarrow} 010\ 111\ 101\ 001 \overset{\text{decoding}}{\longrightarrow} abba.$$

The noise in the channel is flipping some of the 0s into 1s and vice versa. The first *a* is received as a sequence (010) by the receiver. It can either be decoded as *a* or *b*. The two codewords used are (000) for *a* and (111) for *b*. The sequence (010) can originate from (000) in a single flip. Starting from (111), we would need at least two flips. In this sense, the received (010) is closer to the codeword used for *a* than for *b*. Here, the distance is measured as the so-called Hamming distance, which simply counts the number of positions, or "bits," in which the received sequence and the codewords differ.[2]

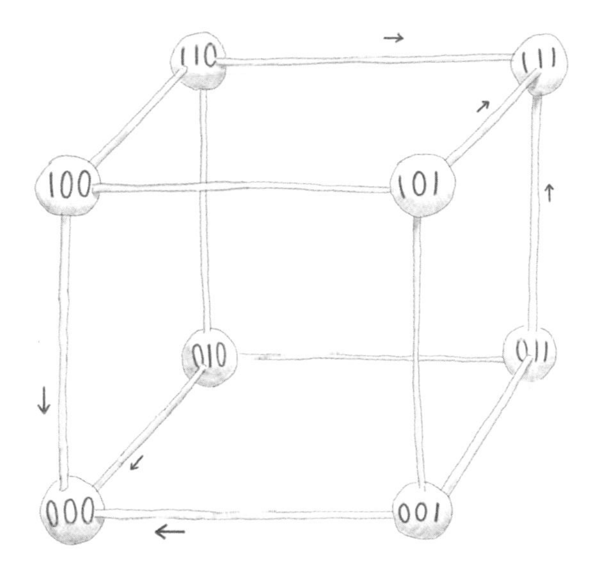

2. The length of codewords is measured in "bits." We say that the encoded message on the previous page has 12 bits. Bits also appear in chapter 5, where they are used for measuring the information content conveyed by the outcome of a random variable.

The cube on the previous page illustrates the decoding. Suppose that we send either a or b. The collection of codewords (also called the "code") contains (000) for a and (111) for b. Each node corresponds to a message that could be received. The nodes (000), (100), (010), and (001) map to the codeword (000) for a, whereas all other nodes map to the codeword (111) for b. If the codewords are transmitted with a maximum error of a single flip, then the decoded sequences will match the true sequences perfectly. In this example, the coding scheme is said to be 1-*error correcting,* as a single flip in the transmission leads to no mistakes. It is also called *perfect,* as all sequences are either codewords themselves or can be generated by a codeword through a single flip. This ensures that after receiving a message, we are never in any doubt as to which codeword to choose in the decoding step. Codes that are 1-error correcting and perfect are called *Hamming codes.*

Decoding the sequence corresponds to associating the received sequence with the closest codeword. If a codeword is disrupted by at most one flip of a bit, then the decoded letter will be identical to the original letter. A mistake will happen if two or all three bits have been flipped by the noise in the channel. If each bit is flipped by the noisy channel with probability $p \in [0, 1/2]$, then each bit is not flipped with probability $(1-p)$. And therefore, each letter is recovered with probability[3]

$$(1-p)^3 + 3(1-p)^2 p.$$

The term $(1-p)^3$ is the probability that the sequence has no flips. The term $3(1-p)^2 p$ is the probability of having only a single flip in the sequence: The probability that only the first bit is flipped is $(1-p)^2 p$, which is also the probability that only the second bit is flipped, and the same for the third. Taking the two

3. The factor 3 in the second term is the binomial coefficient $\binom{3}{1}$; see appendix B.4.

terms together, we obtain the probability of having no flip or just a single flip in the sequence (and hence not making a mistake, as both of these cases will be decoded correctly). The probability of not making a mistake is 97.2% for $p=0.1$, compared with a probability of 90% of receiving the true sequence if we encode the letter a as 1 and transmit b as 0.

1.4 SOLUTION

Let us use a strategy based on the Hamming code that we discussed in Section 1.3. To see how this works, arrange the color sequences on a cube in three dimensions (or n dimensions for n players). All eight possible color sequences of a three-person game correspond to a corner of a cube.

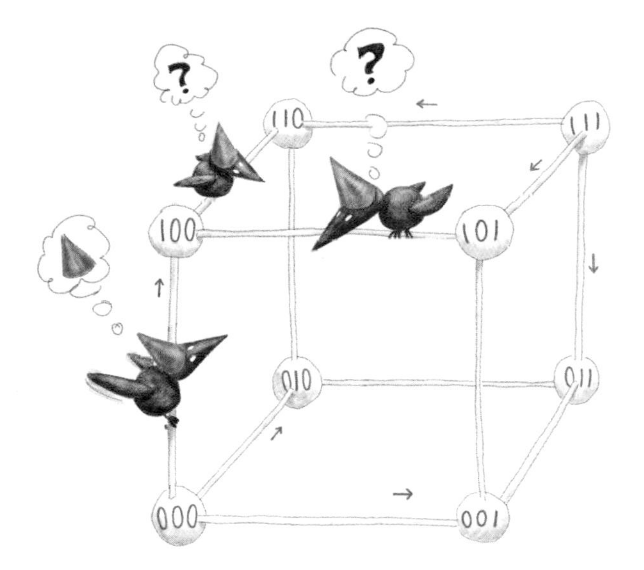

If the true sequence is (100), then the game "sits" at the top-left-front corner of the cube, as shown above. Anwar cannot see that the game is in state (100), of course, as he cannot see the color of his own hat and only knows that the state is either (100) or (000), depending on whether his hat is red or blue. Therefore, we can say that Anwar sits on the edge on the front left, because, for Anwar, this edge connects the two sequences (100) and (000) that are compatible with his observations. Likewise, Charlie can be seen to sit on the top-left edge and Bella on the top front. All edges of the players connect to the true sequence (100), because the true sequence is a possible sequence for all players.

Now, Anwar can decide either to move to the upper node (100) by saying that his hat is red (which it really is), or he can decide, mistakenly, that his hat is blue and thus move to the lower node (000). As a third option, he can decide not to do anything and to answer that he does not know.

Hamming codes allow us to define a successful strategy. If you are sitting on an edge without a codeword as a neighbor, you should say "I don't know." However, if you are sitting next to a codeword, you should choose the sequence on the other side (this is the opposite of the decoding step after transmission over a noisy channel, where you should move toward the codeword, not away from it). We will first illustrate this strategy for three dimensions, $n=3$, where the codewords are (000) and (111). We will consider the case where $n=2^m-1$ dimensions for an arbitrary $m\in\mathbb{N}$ in section 1.5. For the previous example, Anwar (on the front-left edge) is following the arrow on the edge and arrives at the answer "I have a red hat," as the arrow is pointing toward (100) and away from the codeword (000). The other players do not have an arrow on their edges, and they say "I don't know." So, the group wins. In general, if the true sequence is not a codeword of the Hamming code, the group will win, as exactly one player will identify her color correctly, and the rest of the group will not guess a color at all.

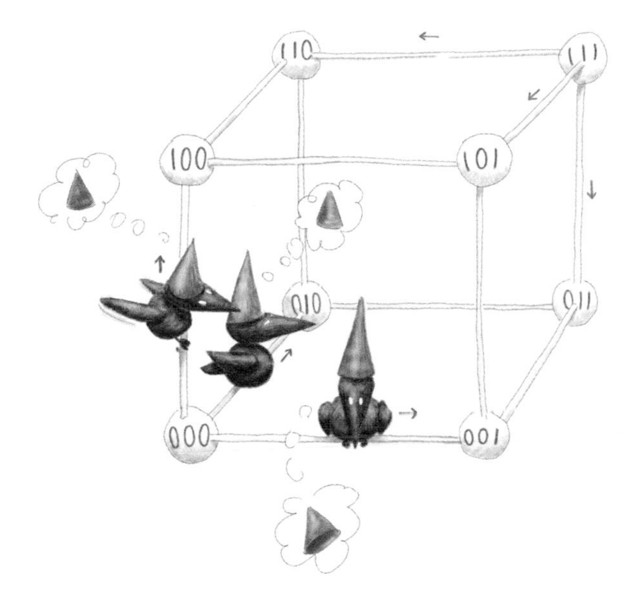

The cube above shows the outcome for the true sequence (000), that is, all players have a blue hat. This sequence coincides with a codeword of the Hamming sequence. All players have an arrow on their edge, and all edges point the wrong way: They will all announce the wrong color (red). The group therefore loses if the true sequence is a codeword.

Therefore, the outcomes are exactly as envisaged when discussing the best possible bound of the failure probability of the group. Either one player produces the correct answer and the others say they do not know, or all players give a wrong answer. A perfect 1-error correcting code guarantees that all states are at most distance 1 from a codeword, and that all neighbors of a codeword are not neighbors of another codeword (the definition of a perfect 1-error correcting code will be explained in more detail in section 1.5). Using the above strategy, the probability of losing is therefore equal to the fraction of codewords among all possible sequences—still assuming that all sequences occur with equal probability. In the example, we have two codewords among eight sequences and hence

$$P(\text{group loses}) = \frac{1}{4}.$$

In the general case of n players, we will see in section 1.5 that the loss probability equals

$$P(\text{group loses}) = \frac{1}{n+1}.$$

This matches the bounds we derived above, and therefore, the win probabilities cannot be improved by using any other strategy.

1.5 HAMMING CODES IN HIGHER DIMENSIONS

In the coding example above, we encoded $\{a, b\}$ by using the codewords $W = \{(000), (111)\}$. We called (000) and (111) sequences, but, more formally, they can be seen as vectors in $\{0, 1\}^3$. Why did we call the code W perfect and 1-error correcting?

The *Hamming distance* $d(x, y)$ for $x, y \in \{0, 1\}^n$ is defined as the number of entries in both vectors that disagree,

$$d(x, y) := \#\{k : x_k \neq y_k\},$$

for example, $d(000110, 010100) = 2$. A ball $B_e(x)$ around $x \in \{0, 1\}^n$ with radius $e > 0$ is the set of all points $y \in \{0, 1\}^n$ that have Hamming distance at most e from x, that is,

$$B_e(x) := \{y \in \{0, 1\}^n : d(x, y) \leq e\}.$$

A code with codewords W is now said to be *e-error correcting* if for all $x, x' \in W$ with $x \neq x'$ and all $y \in B_e(x)$, we have

$$d(x, y) < d(x', y).$$

In words, if y is formed by changing a codeword x in e entries, then y will still be closer to x than to any other codeword $x' \in W$. An e-error correcting code W is said to be *perfect* if

$$\bigcup_{x \in W} B_e(x) = \{0, 1\}^n,$$

that is, the union of all e-balls around codewords is the whole set of sequences $\{0, 1\}^n$. A perfect e-error correcting code is called a *Hamming code*.

The strategy described in the previous section relies on having access to a Hamming code. For $n=3$ players, we showed that $W=\{(000), (111)\}$ can be used, but we could equally well have used $W=\{(110), (001)\}$, for example. Assume that we manage to construct a 1-error correcting perfect code for the general case of n players. Then, for each codeword, there are n sequences that have Hamming distance 1 to this codeword. As the code is perfect, every sequence is either a codeword or has distance 1 to a codeword. For every unsuccessful outcome of the game (the true sequence is a codeword), there are n successful outcomes (the true sequence is not a codeword), which means that the probability of losing is

$$P(\text{group loses}) = \frac{1}{n+1}.$$

The challenge is to construct a perfect 1-error correcting Hamming code for the dimension n that matches the number of players. Currently, this can only be solved when the number of players equals $n=2^m-1$ for some $m \in \mathbb{N}$, for example, $n \in \{3, 7, 15, 31\}$; appendix C.1 shows how such codes can be constructed.

Adversarial Audience

If hat colors are drawn uniformly at random, each of the two sequences all-blue or all-red (which lead to failures of the strategy) occur only with probability 2^{-n}. They might occur more often in practice, however, since the audience might be curious to see what happens in these somewhat special cases. Even worse, if a mischievous audience knows about the Hamming codes, they can always chose a Hamming codeword as the distribution of hats; using the above strategy, this makes the group lose every game. The group, however, can easily protect itself against such opposition by adding a randomly chosen sequence $\{0, 1\}^n$ to all of the codewords and thereby creating a new valid Hamming code (we explain this in more detail in appendix C.1). If the players generate a new Hamming code before each game, they can ensure that the error probability remains the same, no matter what strategy the audience members employ.

1.6 SHORT HISTORY

Todd Ebert introduced what he called the "hat problem" in his PhD thesis at the University of California at Santa Barbara in 1998 [Ebert, 1998]. Many mathematicians tried to solve this problem until Elwyn Berlekamp, a Berkeley math professor, discovered a connection to coding theory and constructed the optimal strategy for the special cases $n = 2^m - 1$ for $m \in \mathbb{N}$. The problem for arbitrary n is still unsolved. Prior to Ebert's game, a similar game was introduced in an article titled "The expressive power of voting polynomials" [Aspnes et al., 1994], in which players cannot take the "I don't know" option, and the objective is to make sure that the majority of players answers correctly. Both games are also described in the *Mathematics Intelligencer* [Buhler, 2002].

1.7 PRACTICAL ADVICE

The figure on the following page shows codewords for two 1-error correcting Hamming codes with $n = 3$ players. The rows in the left half of the figure correspond to the sequences (000) for all blue hats and (111) for all red hats.

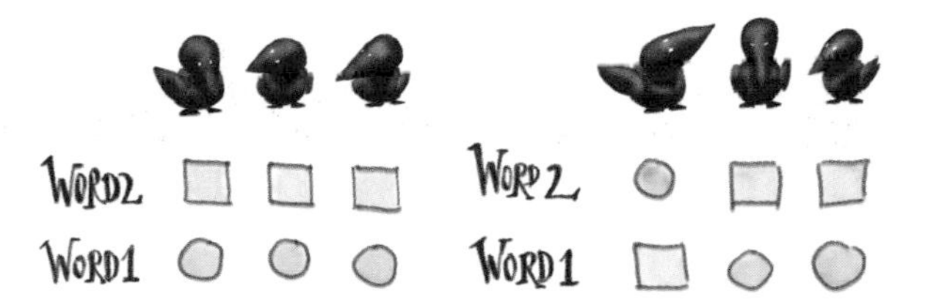

One can add a constant to all codewords, as shown on the right. This yields a new valid Hamming code and can thus be used, too.

The figure above can now be used to derive the optimal action for each player. Suppose we are player 3 in a game with $n=3$ players, and we use the codewords on the left in the figure (all red for word 1 and all blue for word 2). Then we check the colors of players 1 and 2. If they do not match any of the rows for players 1 and 2 in the list of codewords, we answer, "I don't know." This is the case if we see either "blue, red" or "red, blue" for the hat colors of players 1 and 2. If we do see the colors of the other players matching the colors of a codeword (that is either "red, red" or "blue, blue"), then we have to take action by announcing the color that will direct us away from the codeword. If we see "blue, blue," we announce color "red." If we see "red, red," we announce color "blue."

The 16 Hamming codewords of a 1-error correcting perfect Hamming code with $n=7$ players are shown in the figure on the following page. Appendix C.1 shows how to construct the codewords. But it is not too difficult to convince oneself that they indeed form a Hamming code. Each pair of codewords has at least the Hamming distance 3, which means that the code is 1-error correcting. It is a perfect code, as can be shown by a counting argument: Each ball of radius 1 around a codeword contains 8 sequences; in total, the balls contain $16 \cdot 8 = 128$ sequences. Therefore, any of the $2^7 = 128$ sequences of length 7 is contained in one of the balls.

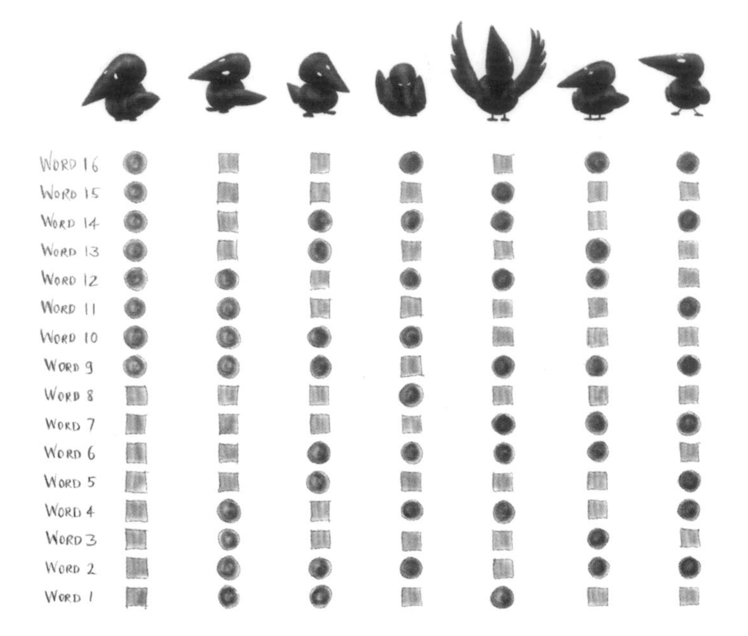

This table can be used analogously to the shorter table shown in the previous figure. Let us say that we are player 3. If we observe for players 1, 2, 4, 5, 6, and 7 the colors "red, red, blue, red, red, and red" (in this order), then the colors match the codeword 9 (with the exception of our own color, which we cannot see). Again, in this scenario, we have to declare in this case the color that will steer us away from the codeword. The matching codeword 9 indicates color "red" for player 3, and therefore we have to declare that, in this case, we have a blue hat. If the colors of the other players do not match any of the 16 codewords, then we answer "I don't know."

2 TWENTY BOXES AND PERMUTATIONS

2.1 THE GAME

Number of players:	10–20
You will need:	1 box for each player; each player has an ID card and a pen (if ID cards are unavailable, each player gets 1 piece of paper)

Let us say we have 20 players. The 20 players stand in a line in alphabetical order. In front of each player is a box with her full name written on it (that is what the pen is for). For simplicity, let us assume there are no players with identical names. The players put their identity cards into their own boxes and are then asked to leave the room. The audience is now allowed to arbitrarily shuffle the ID cards around between the boxes, making sure there is still an ID card inside each box. Then the players are asked back into the room one at a time, where they find the 20 boxes still lined up on the table, as before. The first player called back in has to try to find her personal ID card but is only allowed to check inside half of the boxes, that is, she can look into $20/2 = 10$ of the boxes to find her ID. She is only allowed to look into the box and is not allowed to move any of the ID cards. After the player has checked 10 boxes, she leaves the room, and the next player enters. Every player is only allowed to look into 10 boxes. The players collectively win the game if and only if every one of them finds her ID card. That is, if one or more players do not find their ID cards, all of them fail. After the game has started, the players are neither allowed to communicate with one another nor to interfere with the way the game is set up. Thus, each player will find the boxes and their contents set up in exactly the same way; the players must not move around either the IDs or the boxes.

What is a good search strategy for the players to agree on? How often will they be able to win the game?

Just to make things simple, let us first assume that the audience rearranges the IDs randomly into the boxes. We will discuss later what happens if the audience tries to fool the players.

Some strategies that the players could use are not very clever. Suppose, for example, that every player looks into the first 10 boxes. Working with this strategy, they will always lose the game, since, regardless of the arrangement of IDs, 10 players will not find their IDs.

If, alternatively, every player looks into 10 (independently and uniformly) randomly chosen boxes, then each player recovers their ID with probability 1/2. The chance that all players recover their ID is the product of the individual chances, as the success or failure of a player is independent of the success or failure of any other player with this strategy. For the whole group, the probability of winning is then $p_{\mathrm{win,lower}} = 1/2 \cdot 1/2 \cdot \cdots \cdot 1/2 = 1/2^{20}$. This number is rather tiny: It is comparable to the probability that 2 people sitting silently in front of each other for up to 14 days will choose the same second to shout "Boo!" if both choose a second independently and uniformly within these 14 days. We call $p_{\mathrm{win,lower}}$ a lower bound, since an optimal strategy will have at least the same chance of winning (the same chance only if this strategy were indeed an optimal one, which it is not).

The players could also divide the boxes beforehand and say that players 1–10 check boxes 1–10 and players 11–20 check boxes 11–20. This strategy is obviously better than all players using the same set of 10 boxes, as the players now have a nonzero chance of winning. It also improves on the purely random strategy mentioned above.[1] In fact, it is about 5.6 times better than with the purely random strategy. This is still an almost negligible chance of winning (approximately $3/2^{19}$), but why did it increase in relative terms? The reason is that the division of boxes induces a dependence between the success of individual players. If player 1 is successful (that is, his ID is in one of boxes 1–10, then the chance of, say, the eleventh player to succeed has also increased. Before we know that the first player is successful, the probability that the eleventh player finds his ID under boxes 11–20 is one half. After knowing that the first player has succeeded, the chance of player eleven to succeed is now $10/19 > 1/2$, as there are just

1. One can prove that the chance of winning now equals 1 divided by $\binom{n}{n/2}$, where $n = 20$. According to the so-called Stirling approximation this is roughly $\sqrt{\frac{\pi}{2} n}/(2^n)$ and therefore only slightly larger than $1/(2^n)$, the probability of winning with the purely random strategy.

19 boxes left for his ID instead of the original 20 and each of the 19 boxes left is equally likely to contain his ID. This results in just a very modest boost of the overall success probability, but it is a step in the right direction.

The key to a good strategy is to induce a much stronger dependence between the outcomes, so that if the first player is successful, it will make success of the other players very likely or almost certain (and vice versa, if the first player fails, the others will also very likely fail). The reader should pause here and think once more about the questions presented on page 18.

2.2 HOW WELL CAN A STRATEGY WORK?

Let us now derive an upper bound, so that we can see how often the players are likely to fail, even if they adopt the best-possible strategy. Let us consider the first player to enter the room. She looks into 10 boxes. Her choice of boxes may depend on the agreed-on strategy, of course, but the probability of success does not. No matter which boxes the first person looks into, she will find her ID with probability 1/2. And if the first person fails, the game is certainly lost. (But even if she is lucky, the game is still far from being won.) We therefore have[2]

$$(1/2)^{20} = p_{\text{win,lower}} \leq p_{\text{win,optimal}} \leq p_{\text{win,upper}} = 1/2.$$

Surprisingly, there is a strategy that yields a winning probability that is much larger than the lower bound and that comes close to the upper theoretical limit of the upper bound, so that the group can win almost half of all games.

---

2. Here, $p_{\text{win,lower}}$ is still based on the purely random strategy. We could, in principle, also use the slightly improved version described above, but then $p_{\text{win,lower}}$ would have a more difficult form.

2.3 SOLUTION

The optimal strategy can be described easily. The player with name j first looks into box j (i.e., the box with her own name on it). If she finds her own name in there, she may stop. If she finds another name k (or rather, the ID of person k), say, she then looks into box k. In that box, she may find another name, ℓ, and continues with that box, and so on. The probability of winning with 20 players is about 1/3 and is thus very close to the bound of one-half that we derived in section 2.2.

At first glance, it might not be clear at all why this procedure is successful. The game allows for a natural formulation using the language of permutations. Next we introduce some basic properties of permutations, which will allow us to provide a simple yet thorough analysis of the strategy.

2.4 SOME MATHEMATICS: PERMUTATIONS AND CYCLES

The solution is best understood when we identify the names with the numbers $1, \ldots, 20$. It does not matter which name corresponds to which number, but for simplicity, let us take alphabetical order, in which (by assumption) the boxes are already ordered. So, let us think about the players having numbers between 1 and 20 as names[3] and we will use the terms "names" and "numbers" interchangeably.

Permutations

A *permutation* π of n numbers is a bijective map

$$\pi : \{1, \ldots, n\} \to \{1, \ldots, n\}.$$

Such a map assigns to each number $1, \ldots, n$ one unique number that is again between 1 and n. The term "bijective" means that

3. In fact, this is the setup that is often used when the game is introduced.

each number is "hit" by π exactly once. It is maybe easiest to explain this within the game. Each distribution of IDs into boxes is a permutation: every box j contains an ID (or number) $\pi(j)$, and each ID appears exactly once. That is, for each ID k, there is a unique box j with ID $\pi(j)=k$ in it. Often, it is convenient to write permutations in the following form:

$$\begin{array}{cc|ccccc} j & \text{(box number)} & 1 & 2 & 3 & 4 & 5 \\ \hline \pi(j) & \text{(ID)} & 4 & 5 & 1 & 3 & 2 \end{array}, \qquad (2.1)$$

instead of writing $\pi(1)=4$, $\pi(2)=5$, $\pi(3)=1$, $\pi(4)=3$, and $\pi(5)=2$.

The set of permutations of n numbers is often called S_n and referred to as the *symmetric group*. How many permutations are there? Let us consider the boxes again. For the first box, we can choose from 20 ID cards, for the second box there are 19 possibilities left, and so on. In total, there are thus $20 \cdot 19 \cdot \dots \cdot 2 \cdot 1 = 20!$ different ways of distributing the IDs into the boxes—which is a rather large number:

$$20! = 2432902008176640000.$$

More generally,

$\#S_n = n!$, where $\#S_n$ denotes the number of elements in S_n.

Cycles

There is one more concept that will prove to be helpful when analyzing our strategy. Each permutation can be decomposed into smaller parts, called *cycles*. Suppose that you manually rearrange the ID cards in the boxes. How would you do it? You could take ID card 1 and put it into box 3. You would then take the ID card from that box (ID card 3) and put it into box 4. ID card 4 you would put into the empty box 1. (You would then rearrange the remaining numbers, 2 and 5, in an arbitrary way). You would have thus created a permutation with a cycle that contains the numbers 1, 3, and 4. More formally, a cycle of length m in a permutation π is a sequence $(x_1, \ldots, x_m)$ such that

$$\text{for } j = 1, \ldots, m-1: \pi(x_j) = x_{j+1}, \quad \text{and} \quad \pi(x_m) = x_1.$$

You can "walk around a cycle," and when you try to "walk around the permutation," you are always stuck in one of its cycles. For example, the permutation described by equation (2.1) has one cycle of length 3: $(1, 4, 3)$. This is because the permutation satisfies $\pi(1) = 4$, $\pi(4) = 3$, and $\pi(3) = 1$. The other cycle of equation (2.1) has length 2: it is the cycle $(2, 5)$.

2.5 UNDERSTANDING THE SOLUTION

Using the notation introduced above, let us describe the distribution of numbers into 20 boxes by a permutation $\pi \in S_{20}$: Box 1 contains number (or ID) $\pi(1)$, box 2 contains number $\pi(2)$, and so forth:

box number	1	2	3	$\cdots$	20
ID	$\pi(1)$	$\pi(2)$	$\pi(3)$	$\cdots$	$\pi(20)$

Remember that permutations are bijective (i.e., all the boxes contain different IDs). For now, we model the distribution of the IDs by randomly choosing a permutation π from a uniform distribution on S_n. That is, when the game is set up, each permutation can be chosen with probability

$$\frac{1}{20!}.$$

How can we interpret the proposed strategy in terms of permutations? Player j looks into box number j and finds the ID with number $\pi(j)$. She then looks into box number $\pi(j)$, finds ID $\pi(\pi(j))$, and so on. The key step is to understand when the game is lost. Let us therefore consider a small example with 10 players, where each player is allowed to look into 5 boxes, say:

box number	1	2	3	4	5	6	7	8	9	10
ID	5	4	8	6	1	9	3	7	2	10

The distribution of IDs can also be visualised as a graph, where each node corresponds to a box. Each box has a directed edge to another box. The value of the ID hidden under the box determines which box the arrow points to, and it is the next box the player would check if following the strategy.

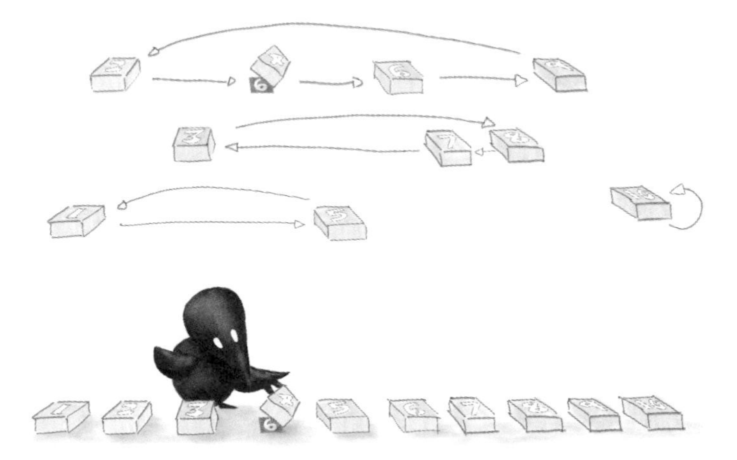

Let us consider the game specifically from player 4's point of view (see the above figure). She starts with box 4 and finds ID number 6 underneath. She then checks box number 6 and finds number 9. In box 9, she finds number 2, and in box 2, finally, she finds her own ID, the number 4. Mathematically, she has walked around a cycle of the permutation!

The strategy guarantees that each player starts in "her own" cycle, the cycle that contains her ID. Moreover, her ID will always be under the last box in the cycle, as this box is "pointing" toward the box that the player started from. Player 4 starts at box 4, and box 2 is pointing toward her starting box, as it contains her ID with the number 4. In this way, each player walks around her own cycle and continues until she either finds her number, or the maximum number of attempts is reached. In the example above, not only does player 4 find her number, but so do all other players. This is not always the case. In the following example, the strategy fails:

box number	1	2	3	4	5	6	7	8	9	10
ID	3	4	8	6	9	10	2	7	5	1

Visualized as a graph, the situation is now as shown in the following figure.

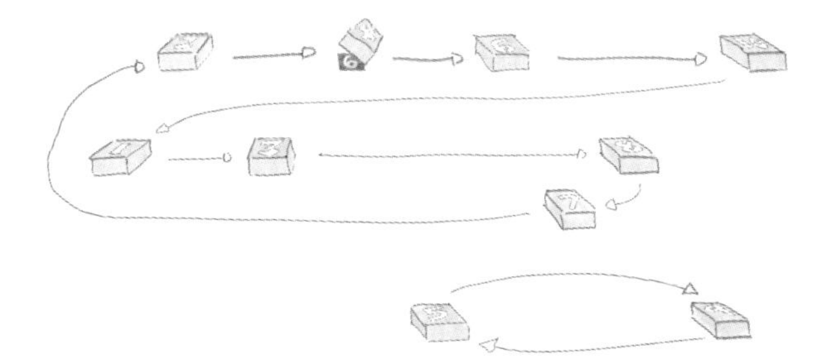

In this example, player 4 fails to find her number; she has to stop after 5 attempts, that is, after looking into box 3. What is the crucial difference between the two examples? The second permutation contains a cycle of length 8: $(4, 6, 10, 1, 3, 8, 7, 2)$. The fact that it is longer than the allowed number of attempts, 5, leads to a disastrous loss of the group: Not only does player 4 remain out of luck, but so do players 2, 6, 10, 1, 3, 8, and 7. These are all the players with a number in the same cycle, and none of them is able to find the correct ID. Players 5 and 9, in contrast, do find their own numbers after looking under just two boxes. If, as in the first example, all cycles are of short length, then all players are guaranteed to find their number.

This finding brings us much closer to understanding the solution. Using the proposed strategy, the game is lost if, and only if, the permutation representing the distribution of ID cards contains a cycle that is longer than the maximal number of attempts. If all cycles are small (in our example with 20 players, that is, length 10 or smaller) the players will eventually find their own ID in 10 attempts or fewer. If there is a large cycle (length 11 or larger), there will be players (at least 11 of them) who will not find their number, because they can only check 10 boxes and therefore cannot complete the walk around their own, long cycle. To compute the probability of losing, we therefore need to answer the question:

What is the probability that the permutation has a long cycle?

We will come back to this question soon, but first, we point out a similarity between the box game and the hat game presented in chapter 1: The strategy makes the players' outcomes dependent on one another.

Relation to the Hamming Code Solution

This game is similar to the game described in chapter 1, in that the players win and lose together. There, we discussed a solution that collects all wrong answers in single instances of the game and distributes the correct answers over as many games as possible. In the box game discussed here, we observe a similar effect: If there is a long cycle that prevents one player from finding her ID card, many other players will be unsuccessful, too. In this sense, the outcome of the individual players are no longer independent. In many games that are won and lost jointly, the key idea is to introduce a dependence between the individual successes. The challenge is usually finding out how to create that dependence.

Computing the Probability of Winning

To compute the probability that a permutation has a long cycle, let us count the number of permutations in S_{20} that give rise to a cycle of length k with $11 \leq k \leq 20$. There are $\binom{20}{k}$ possibilities of choosing the k numbers that are involved in the cycle. Here,

$$\binom{20}{k} := \frac{20!}{k!(20-k)!}$$

is called the "binomial coefficient" and describes the number of different possibilities to choose a set of k numbers out of a set of 20 numbers; see "What Is ... a Binomial Coefficient?" (appendix B.4). These k numbers can be sorted in $(k-1)!$ different ways to form a cycle of length k (because we can always start with the lowest number). The cycle leaves us with $20-k$ elements for the rest of the permutation, which can be arranged in $(20-k)!$ ways. Thus, there are $\binom{20}{k}(k-1)!(20-k)!$ permutations with a cycle of length k. To summarize, the number of permutations with a cycle of length 11 or larger equals

$$\sum_{k=11}^{20} \binom{20}{k}(k-1)!(20-k)! = \sum_{k=11}^{20} \frac{20!}{k}.$$

Importantly, we did not count any permutation twice: If we form a cycle of length 15, for example, it is impossible to obtain, in the remaining numbers, another cycle of length 11 that we might have counted before. The probability that a randomly chosen permutation has a cycle of length 11 or larger therefore becomes

$$\frac{\sum_{k=11}^{20} \frac{20!}{k}}{20!} = \sum_{k=11}^{20} \frac{1}{k} \approx 0.67.$$

The probability of winning equals the probability that there is no large cycle, and therefore we obtain

$$p_{\text{win, optimal}} \approx 1 - 0.67 = 0.33,$$

which is surprisingly close to the upper bound of 0.5. Strictly speaking, we do not know yet that this strategy is really optimal, but we will come back to that question later.

More Players

If we keep the rule that players are allowed to look into half of the boxes, what happens to our strategy if we increase the number of players (while keeping it even, say)? Surprisingly, the probability of winning does not change much! The analysis is completely analogous to the one already presented. The probability of losing the game for n players who are allowed to look into $n/2$ boxes equals

$$\frac{\sum_{k=n/2+1}^{n} \frac{n!}{k}}{n!} = \sum_{k=n/2+1}^{n} \frac{1}{k}, \tag{2.2}$$

still, assuming that n is even. The number on the right side of equation (2.2) is not so easy to compute directly, but we can use integrals to approximate it. (If you have not seen integrals before, just skip to the final result of this paragraph.) By drawing the graph of the function $x \mapsto 1/x$, it is not difficult to see that we have

$$\int_{n/2+1}^{n+1} \frac{1}{x}\, dx \le \sum_{k=n/2+1}^{n} \frac{1}{k} \le \int_{n/2}^{n} \frac{1}{x}\, dx.$$

But these inequalities imply that the probability of losing is (slightly smaller than)

$$\log(n) - \log(n/2) = \log(2) \approx 0.6931472$$

and converges for $n \to \infty$ to log(2). (Here, we consider the "natural" logarithm to the base of the Eulerian number; see "What Is … an Exponential Function?" in appendix B.3.) The probability of winning therefore converges from above to

$$1 - \log(2) \approx 0.3068528.$$

Optimality

We mentioned that the above strategy is optimal. Let us now see why this is indeed the case. Remember that we are assuming that the IDs are distributed randomly. First, let us alter the game: The players enter the room according to their number, starting with 1, then 2, and so forth. Each player is allowed to look into as many boxes as she likes until she finds her ID. The game is won if each player requires $n/2$ attempts or fewer. In terms of winning and losing, this game is equivalent to the original version, so we call this slightly adapted version the "original game" below. Now, let us change the game even further, so that we create a new game: All players are present all the time, and after the first player has found her ID, all players whose ID has been revealed by the first player are out of the game, together with the corresponding boxes. Then the player with the smallest number among those whose ID has not yet been found starts to check some of the remaining boxes until she finds her number, and so forth. As before, the game is won if none of the players used more than $n/2$ attempts. There are two key observations:

1. Any strategy in the original game can be applied to the new game and will yield (at least) the same probability of winning in the new game.

Let us now assume that, for the new game, we construct a protocol by writing down the order in which the IDs have been revealed (ignoring the information regarding which player revealed it). This protocol order lets us reconstruct what happened during the game. For example, the protocol

$$(19, 8, 1, 10, 3, 9, 20, 5, 18, 11, 12, 7, 6, 4, 15, 2, 17, 16, 13, 14)$$

tells us that the first player used 3 attempts, the second player opened 13 boxes, player number 13 opened 3 boxes, and, finally, player 14 opened the last box. This order from the protocol can again be interpreted as a permutation: one that contains the cycles $(19, 8, 1)$, $(10, 3, 9, 20, 5, 18, 11, 12, 7, 6, 4, 15, 2)$, $(17, 16, 13)$, and (14). We see that the new game is won if and only if that permutation does not contain a cycle that is longer than 10 (if it did, one of the players would have looked into more than $n/2$ boxes).

2. The protocol describes a permutation that is chosen (uniformly) at random.

Why is this? The protocol order clearly depends on the team's strategy, but since we started with a random permutation in the first place, the protocol order will be a random permutation of the numbers 1, ..., 20, independently of what strategy the players have used. The permutation induced by the protocol is therefore also chosen uniformly at random (this is because there is a one-to-one correspondence between protocol orders and protocol permutations).

In summary, the probability of winning the new game is independent of the teams' strategy and equals the probability that a random permutation has no cycle that is longer than $n/2$. Finally, observation 1 tells us that this probability is larger than that of success for any strategy in the original game. This proves that the above strategy is indeed optimal for the original game.

Playing against an Adversary

If the audience knows about the strategy, they could try to set a trap for the players. Instead of choosing a random permutation, they could deliberately choose one with a long cycle. If the players follow their own strategy, they will definitely lose the game. However, the players can protect themselves from an audience that is working against them: They can draw a random permutation τ before starting to play, and then they can reinterpret each ID number k, including their name and the numbers that they find as $\tau(k)$ (but not the box labels). Again, they will end up with the above-computed winning probability.

2.6 SHORT HISTORY

The game was introduced by computer scientists Anna Gál and Peter Bro Miltersen in 2003 [Gál and Miltersen, 2003]. The game was used to illustrate the proof of one of their theorems, and they won a best-paper award for their conference paper. In their original version of the game, each player's ID card is either red or blue (which the players do not know), and every player has to guess the color of their ID card while turning over only half of the boxes. The probability of winning is marginally higher in this game, as players can of course just guess when they have not found their card. Some further variations of the game can be found in a paper by Navin Goyal and Michael Saks [Goyal and Saks, 2005], while a nicely written overview of the history can be found in an article by Eugene Curtin and Max Warshauer in [Curtin and Warshauer, 2006].

2.7 PRACTICAL ADVICE

When playing the game, one can consider increasing the proportion of boxes that the players are allowed to check. This yields a larger probability of success, which might increase the players' motivation. For example, when allowing the players to look into 3/4 of the boxes, the probability of losing equals

$$\sum_{k=3n/4+1}^{n} \frac{1}{k},$$

and therefore the probability of winning converges to

$$1-(\log(n)-\log(3n/4)) = 1-\log(4/3) \approx 0.7123179.$$

This game is more likely to yield a successful run, especially if the game is repeated a few times.

3 THE DOVETAIL TRICK AND RISING SEQUENCES

3.1 THE TRICK

Number of players:	1 magician who knows about the card trick; 1 audience member who is able to perform a riffle shuffle
You will need:	card deck with 52 different cards

A magician hands over a (sorted) deck of cards to a person in the audience. The magician turns away and then asks the contestant to riffle shuffle the deck three times (an example of a riffle shuffle is shown in the picture on page 35). She further asks the contestant to cut the deck at a place of his choice: that is, she asks him to split the deck into two parts and place the top part below the bottom part. The contestant then looks at and memorizes the new top card and places it somewhere in the middle of the deck. Finally, he is again allowed to cut the deck. The magician looks at the deck, studies it for a few seconds, and then picks out one card by elevating it, as shown in the figure on the following page.[1] There is no sign that the magician was secretly

1. The figures in this chapter containing playing cards were produced using the package, "pst-poker," which is maintained by Herbert Voß.

overlooking the scene, and nobody from the audience has given a hint.

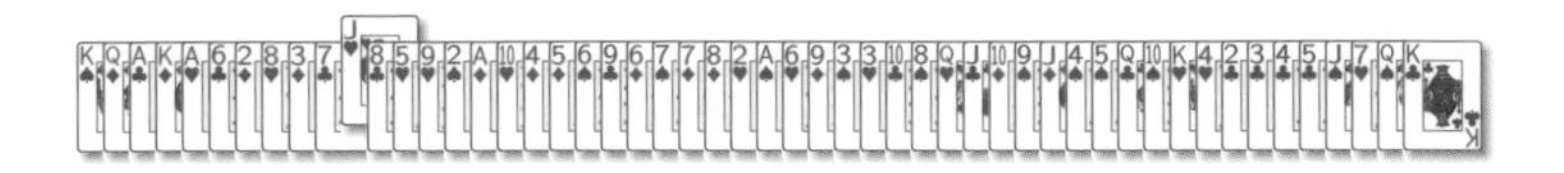

Indeed, the Jack of hearts was the card the contestant looked at before!

How did the magician find the card?

The trick is based on mathematical insight and not, for example, on card manipulation or cooperation with a member of the audience. Just seeing the sequence of cards as in the example above (but without the elevated card, of course) was enough for the magician to pick out the correct card.

Suppose that instead of three riffle shuffles, the audience member performs a "perfect" shuffle. Then, any order of cards would be equally likely, and it becomes impossible to find the chosen card. The trick thus has to make use of the fact that the cards were not perfectly shuffled. In fact, the magician starts with an ordered set of cards and asks for only three consecutive riffle shuffles. Even though the resulting ordering of cards looks random to the untrained observer, there is a lot of nonrandom structure in the deck. In the remainder of this chapter, we will learn about how to describe that structure and how to exploit it for the trick. You might want to pause here and think about what lets us distinguish the deck shown above from a perfectly shuffled deck of cards. How did the magician find the card?

3.2 RIFFLE SHUFFLING CARDS

In this chapter, we talk about playing cards. The following list contains some useful terms:

suit	sometimes also called the *color* of a card; we use so-called French playing cards with four suits: *diamonds* (♢), *hearts* (♡), *spades* (♠), and *clubs* (♣).
deck	a collection of cards; we use 52 playing cards that contain the two of hearts and the king of spades, for example, but no jokers.
sorted deck	a deck in the order: two of diamonds,..., king of clubs, ace of clubs.
shuffle	putting the cards in a new, at best completely random order; there are different shuffling techniques; we will concentrate on the so-called *riffle shuffle*.
cut	the deck is split into two parts, the top part is put on the bottom, such that the former bottom part becomes the new top part (often applied after a shuffle).

The trick described above works only with the riffle (or dovetail) shuffle, a shuffling technique that is very common among card players. The deck is first split at a random position and the two piles are then joined together as shown in the following picture.

It takes a bit of practice before one can perform the shuffle smoothly. When used for card games, the procedure is usually repeated a couple of times. The number of repetitions, however, varies among different players. In our experience, when you shuffle only once or twice, the other players might complain or might simply become suspicious. Performing three or four shuffles, however, is usually considered adequate.

Let us recall the purpose of shuffling. Suppose we play a card game, in which, after some time, the whole deck lies openly on the table. Before we redistribute the cards to the players and start another round of the game, we shuffle the deck. If this shuffle is done well, the deck of cards is put into an entirely random sequence. This ensures that none of the players can benefit from memorizing the order in which the cards have been played during the last round. This is particularly important for casinos. If played optimally, blackjack, for example, is close to being a fair game. The player will, on average, lose only 0.6% of his bet to the bank (this average varies slightly, depending on the precise set of rules). This, however, is only true if the deck of cards is in an entirely random order. If the shuffle has not been done well, and there is a bit of structure left in the deck, then skilled players may be able to exploit the structure, and the casino will lose money in the longrun. This is why casinos became interested in an answer to the question:

What is a good number of riffle shuffles?

The magician and mathematician Persi Diaconis famously argued that a reasonable answer to this question is "around seven times." In section 3.6, we will examine the reasoning behind this answer. In fact, the question is very much related to the dovetail trick. The key reason that the dovetail trick works is that performing only three riffle shuffles is not a good shuffle:

The deck is far from being in a random order and still contains a lot of structure from the previous order.

Let us now translate this problem into the language of mathematics. Some insight on permutations (section 3.3) will help explain how the dovetail trick works (section 3.4). To answer the question about a good number of shuffles, we have to work a bit harder and finally present Diaconis's famous answer in section 3.6.

3.3 SOME MATHEMATICS: PERMUTATIONS

You have no doubt already guessed that we are going to use the language of permutations seen in section 2.4 to study shuffles. But how, exactly, do we model a shuffle? Suppose that we have a deck with 52 cards. We first encode the cards by numbers $\{1, 2, \ldots, 52\}$, as shown in the following figure:

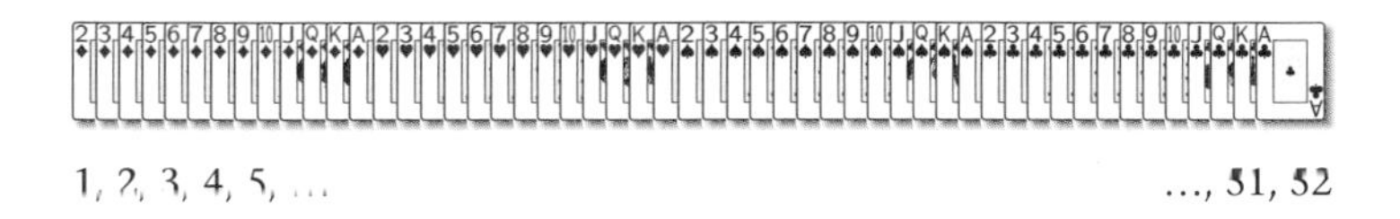

In the remainder of this chapter, we use the numbers between 1 and 52 and the front faces of the playing cards interchangeably. Before the shuffle, we find the deck in a certain ordering: π_{before}, say. This can be really any ordering, but for simplicity, let us assume that the cards are initially in the original order. After we riffle shuffle the deck, it is (hopefully) in a different order, for example,

$$\pi_{\text{before}} = (1, 2, 3, 4, 5, 6, \ldots, 51, 52),$$

$$\pi_{\text{after}} = (15, 3, 22, 16, 4, \ldots, 9, 21).$$

Each of these two orderings corresponds to a permutation of the numbers $\{1, \ldots, 52\}$, and we can write $\pi_{\text{before}}, \pi_{\text{after}} \in S_{52}$, as we did in section 2.4. The number of possible permutations (or orderings) π_{after} equals $52! = 52 \cdot 51 \cdot \cdots \cdot 1$, a number with 67 digits that does not even fit into one line if we write it down:

$$52! = 80,658,175,170,943,878,571,660,636,856,403,766,$$
$$975,289,505,440,883,277,824,000,000,000,000.$$

In fact, there are more orderings of a card deck than there are atoms in our solar system.[2] A shuffle is perfectly random if each of these many, many orderings is obtained with the same probability. That is, after performing it, we do not know anything about the order of the cards.

The following key observation will help us solve both strategy questions posted in this chapter. After performing a single riffle shuffle, the vast majority of orders are impossible to obtain. We will show later that we can "only" reach

$$4,503,599,627,370,444$$

different orders, many of which are still very unlikely. In particular, this means that after one riffle shuffle, the deck of cards is far from being "perfectly randomly shuffled." The concept of rising sequences will help us to understand this better.

Rising Sequences

A *rising sequence* is a sequence of cards that is constructed by cycling through the deck and for each card, looking for its successor in the sorted deck. The concept is probably best understood by looking at an example. Assume that after one riffle shuffle, we obtain the following deck:

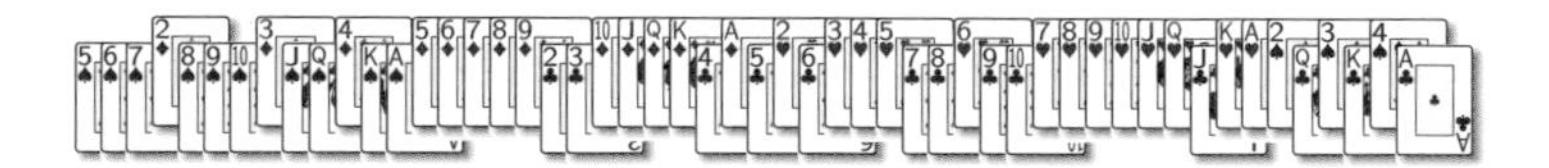

2. This is mind blowing, is it not?

It contains two rising sequences:

(5 of spades, 6 of spades, 7 of spades, ..., ace of clubs) and
(2 of diamonds, 3 of diamonds, 4 of diamonds, ..., 4 of spades),

or, in terms of numbers:

$$(30, 31, 32, 33, \ldots, 52) \text{ and } (1, 2, 3, \ldots, 29).$$

To find all rising sequences (in any deck), one starts with the leftmost card that has not yet been visited and then always searches for the current card's successor. If we reach the end of the deck, we start again at the leftmost card that has not yet been visited, begin a new rising sequence, and iterate. Since a riffle shuffle is the combination of (half) a cut and an interleaving process, it is not surprising that the number of rising sequences in the above example is 2. For a completely randomly shuffled deck, however, having only 2 rising sequences is very unlikely: we might have up to 51 rising sequences (we explain below why it is not 52), and the probability of obtaining at most 2 rising sequences is much less than 0.000000001%. In fact, it is less than 10^{-50}%.

If we riffle shuffle twice, there are at most 4 rising sequences; if we riffle shuffle 3 times, there are at most 8 rising sequences, and so forth. In general, we have the following formula:

m riffle shuffles yield up to 2^m (or fewer) rising sequences.

If you would like, you are welcome to take a break and try to convince yourself that this statement is correct. But we will, in any case, come back to this question in section 3.5.

There is an important detail to consider when counting rising sequences: You should think of the two of diamonds as the successor of the ace of clubs. Mathematically speaking, this means that $52+1=1$. This is sometimes called, "computing modulo 52," a technique that we will see again in chapter 4. The fact that we "keep going," even after reaching the ace of clubs, explains why there can be at most 51 rising sequences and furthermore makes our lives easier when dealing with cuts.

Cuts

Cuts play a surprisingly small role when analyzing riffle shuffles. If you cut a deck, the order of the cards is not really altered in any important way. For instance—thanks to our rule that the two of diamonds is the successor of the ace of clubs—cutting the deck after a riffle shuffle does not change anything about the number of rising sequences. Let us consider why this is the case. After riffle shuffling a sorted deck, one of the rising sequences usually[3] starts with the two of diamonds, as in the example shown on page 38. If the deck is now cut, this is not true anymore—the starting points of the rising sequences change. But, importantly, the number remains the same. The only change is that the two of diamonds has lost its special role as the "starting point" of the deck.

In many card games, the dealer changes with each round, so as to evenly distribute the burden of shuffling. It is common that after shuffling the cards, the dealer asks another player to cut the deck. This does not have a strong influence on the hands that are dealt out. In fact, if the dealer distributes one card to each player at a time, the hands remain exactly the same. The only thing that changes is which player will get which hand. If a skilled dealer decides to cheat and creates an exceptionally good hand using a special shuffling technique, the cutting process ensures that the dealer has no control over which player ends up with the winning hand.

---

3. This is not the case only if the riffle shuffle is slightly degenerate: After cutting the deck, the whole bottom part is put on the top part. This results in a deck with only one rising sequence, which does not necessarily start with the two of diamonds.

3.4 SOLUTION

After 3 riffle shuffles, we can expect up to $8 = 2^3$ rising sequences (we will come back to why this is true in section 3.5). The deck in the example displayed on page 34, however, contains 9 rising sequences. In the following picture, we have rearranged the same deck to highlight the rising sequences:

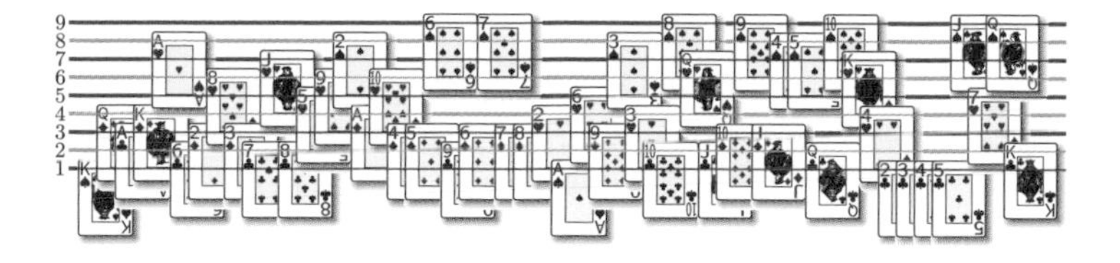

The first 3 rising sequences start with the king of spades, six of clubs, and ace of clubs, respectively. Where does the additional, ninth sequence come from? The sought-after card that the contestant has put somewhere in the middle of the deck is responsible for creating the extra rising sequence! But which one is it? If you look at the example closely, there is only a single card in the example that you can move to a position that would reduce the number of rising sequences to 8. (The figure on page 34 reveals that card.) In general, the dovetail trick does not always work, but here, the magician is 100% certain about the true card and could even incorporate her confidence into the performance.

The card trick does not work, for example, if the top card (the one we are trying to find) is moved only a tiny bit and does not create an extra rising sequence. Or, alternatively, there may be an extra rising sequence, but there is more than one way of resolving it. In the example above, this happens if the nine of hearts and the jack of hearts switch positions. In such a scenario, the magician may decide to stop the trick and somehow talk herself out of it. But it is also possible to still choose a card. In practice, one can assign a score to each card that measures how well it fits into the collection of rising sequences. It is useful, for example, to count for each card the distances to its two neighbors and add those distances together. That is, you start at a card (nine of spades, say) and count the number of steps to the right until you meet its successor, the ten of spades. In the above example, this is 5 (you might find it easier to look at the picture on page 34). To reach its predecessor, the eight of spades, you need 4 steps to the left. The nine of spades therefore receives the score 9. If you reach one end of the deck before finding the card you are looking for, you simply continue at the other end. For example, the six of diamonds receives the score 5, the ace of hearts the score 24, and the jack of hearts 69, which turns out to be the highest score in this deck. Only cards with a score higher than 52 have the potential to reduce the number of rising sequences by one. And if there is only one such card, such as in our example, this must have been the contestant's card.

Below, we show two more examples of decks that the magician might face. We encourage you to play the role of the magician now and find the card that has been put back into the deck.

Here is a small hint for this first example: The correct card has the highest score of 54; the second highest score is 38. In practice, computing the exact score may take some time. You can use this second example to explore your own ways of detecting the card as quickly as possible. Solutions to both examples can be found on the bottom of page 133.

Using the scoring technique described above, the success probability of the trick is about 84% and even increases to about 94% if the magician is allowed to have two guesses (corresponding to the two highest-scoring cards).

3.5 MORE MATHEMATICS: SHUFFLING DISTRIBUTIONS

When shuffling cards, we have started from an order π_{before} and obtained the order π_{after}, where, for simplicity, we still assume that we start with

$$\pi_{\text{before}} = (1, 2, 3, 4, 5, 6, \ldots, 51, 52).$$

Most shuffles have a random component, so even if we perform the same type of shuffle (such as a single riffle shuffle), to two sorted decks, the two resulting orderings π_{after} will in general not be the same. We therefore describe the type of shuffle by a probability distribution over permutations, which we denote by the letter Q. That is, a shuffling type is characterized by assigning a probability to each possible order π_{after}:

$$\pi_{\text{after}} \mapsto Q(\pi_{\text{after}}).$$

Because the function Q describes probabilities, we have the following properties (sometimes called *Kolmogorov's axioms*): All new orderings π_{after} satisfy $0 \le Q(\pi_{\text{after}}) \le 1$, and the sum of probabilities over all possible new orderings equals one, that is, $\sum_{\pi_{\text{after}} \in S_{52}} Q(\pi_{\text{after}}) = 1$. Appendix B.5, "What Is ... a Probability?" contains more details on probabilities.

As an example, consider the following very simple shuffling technique that we name a "half-shuffle." The shuffler randomly chooses a number from $\{24, 25, 26\}$, each of them with probability $1/3$, and then cuts the deck after the card with that number. This technique would most likely yield some complaints from your fellow players, but for this shuffle, $Q = Q_{\text{half}}$ has a simple form:

$$Q_{\text{half}}(\pi_{\text{after}}) = \begin{cases} 1/3 & \text{if } \pi_{\text{after}} = (25, 26, \ldots, 51, 52, \ldots, 23, 24) \\ 1/3 & \text{if } \pi_{\text{after}} = (26, 27, \ldots, 52, 1, \ldots, 24, 25) \\ 1/3 & \text{if } \pi_{\text{after}} = (27, 28, \ldots, 1, 2, \ldots, 25, 26) \\ 0 & \text{otherwise.} \end{cases} \tag{3.1}$$

We can now formalize what we mean by a "perfect shuffle:" All orders are equally likely, that is,

$$\text{for all } \pi_{\text{after}},\ Q_{\text{perfect}}(\pi_{\text{after}}) = \frac{1}{52!}. \tag{3.2}$$

In the case of riffle shuffles, the function Q is rather complicated. We expect that the probability depends on the number a of riffle shuffles that are performed. We therefore need to find a probability distribution $\pi_{\text{after}} \mapsto Q_a(\pi_{\text{after}})$ that is a good model for an a-shuffle (i.e., for performing a riffle shuffles in a row). We discuss such a model below.

Once we have introduced that model with its distribution Q_a, we can then analyze the "goodness" of these shuffles. As often, there is no universal definition of goodness, but since a perfect shuffle is represented by a uniform distribution over all orders, we will define goodness by the difference between the distribution Q_a and a uniform distribution Q_{perfect}. We discuss this in detail in section 3.6.

A Model for Riffle Shuffles

The so-called Gilbert-Shannon-Reeds model attempts to yield a mathematical description of what happens in a riffle shuffle. There are several formulations of this model, all of which are equivalent. Thus, all of these descriptions yield the same probability distribution Q_a. We will start by describing three equivalent models for $a = 1$ and introduce the general case $a \geq 1$ later. In fact, let us introduce an ℓ-shuffle and use the following correspondence:

1 riffle shuffle	$\leftrightarrow$	2-shuffle, $\ell = 2$
2 riffle shuffles	$\leftrightarrow$	4-shuffle, $\ell = 4$
3 riffle shuffles	$\leftrightarrow$	8-shuffle, $\ell = 8$
a riffle shuffles	$\leftrightarrow$	2^a-shuffle, $\ell = 2^a$.

Let us start by describing 1 riffle shuffle, that is, a 2-shuffle.

(A.2) This description is close to the shuffle we perform in practice: First, the deck is split into two parts—this is sometimes called the "half-cut." The size of the first half is around half of the deck. More precisely, the probability that the top part contains k cards equals[4]

$$\binom{52}{k} \frac{1}{2^{52}}.$$

Here, $\binom{52}{k}$ is the *binomial coefficient* (see appendix B.4); it is largest when $k = 26$ and becomes relatively tiny when k is much smaller or larger than 26. This means that the half-cut of the riffle shuffle usually happens somewhere around the middle of the deck.

Afterward, the two parts of the deck are put together again. Let us assume that, when doing so, the dealer is more likely to mix in cards from the larger of the two piles, which means that both piles run out of cards at approximately the same time (see the figure below). Specifically, the two parts are merged together by the following rule: For any given step, we have x cards in the left and y cards in the right pile, say. The next card comes with probability $x/(x+y)$ from the left and with probability $y/(x+y)$ from the right part.

4. We say that the number k follows a *binomial distribution* with parameters $n = 52$ and $p = 0.5$.

(B.2) An alternative (and equivalent) description is to sample the rising sequences directly. To do so, we draw 52 numbers, each of them being either 1 or 2, with equal probability, and the results do not influence each other (the numbers are said to be drawn independently and are identically distributed or, for short, i.i.d.). The resulting sequence, for example,

2221222122122111112211112121211122122111112

11121212,

encodes π_{after}. Specifically, the sequence says that the first three cards of π_{after} belong to the second rising sequence, the fourth card to the first rising sequence, and so forth. The sequence contains 29 ones, that is, the first card in the second rising sequence is card number 30 (or the five of spades). In fact, the sequence encodes exactly the deck you see on page 38.

Note that there can be a few sequences that encode the same deck. For example, a sequence with only ones and three twos at the end encodes the same deck as a sequence with only ones and ending with a single two.

(C.2) The last description is rather abstract, but very useful for proving mathematical properties. We start with 52 numbers drawn i.i.d. from a uniform distribution[5] on the interval $[0, 1]$. We then attach the labels $1, 2, \ldots, 52$ to the points, the smallest number receives the label 1, the largest number, the label 52. We then map these points according to

$$x \mapsto 2x \bmod 1.$$

Here, "mod 1" should be understood as removing the digits in front of the comma (e.g., $1.435 \bmod 1 = 0.435$). The resulting (permuted) sequence of the labels is then the new order of the deck.

One can show that all of these descriptions are indeed mathematically equivalent. But do they provide a good mathematical description for what is happening in reality when people perform riffle shuffles? This is a typical question in the field of statistics. Suppose that you are playing backgammon against someone who would like to use one of her own dice. She assures you that the die is fair, and you ask her to throw it a few times. You might be fine with the result 6, 2, 6, 1, 4, 3, but if she throws 6, 1, 6, 6, 6, 6, you might start to get suspicious. Where do we draw the line? In statistics, we are given a probabilistic model about some data-generating process; we then analyze whether for a certain data set, it is plausible that the set came from that model or whether the data are too unlikely, given the model

5. This means that for each of the 52 numbers, the probability of falling into an interval $[a, b]$ with $0 \le a \le b \le 1$ equals $b - a$. Furthermore, the values of some of these numbers do not provide us with any information about where the remaining numbers will fall.

assumptions. Such an analysis has been performed with the above shuffling model: People were asked to perform riffle shuffles, and the outcomes have been considered plausible under the model.

The above models generalize to the case $\ell \geq 1$. You may think about $\ell = 2^a$, which represents the case of a riffle shuffles (e.g., $\ell = 2^3 = 8$ corresponds to performing 3 riffle shuffles). We need only a slight adaptation from the descriptions above.

(A.ℓ) We cut the deck into ℓ piles. The probability that the piles have sizes $k_1, \ldots, k_\ell$ equals[6]

$$\binom{52}{k_1 \cdots k_\ell} \frac{1}{\ell^{52}},$$

where $\binom{52}{k_1 \cdots k_\ell}$ is the *multinomial coefficient*, explained in appendix B.4. The piles are joined according to a similar rule as in description (A.2): The probability that a card is "chosen" from pile j is proportional to the number of remaining cards in pile j.

(B.ℓ) We choose 52 numbers uniformly and i.i.d. from $\{1, \ldots, \ell\}$. If number k equals 3, say, it means that the kth card in π_{after} is in the third rising sequence. The sequence

1434823632725684633923933415384297239388297411
119592

encodes the deck that is shown on page 41. (In fact, the deck shown on that page is an example of the card trick, and as such, is the result of an 8-shuffle, a cut, a replacement of the top card, and another cut. But the sequence encoding shows that the deck could have also been the result of a 9-shuffle.)

6. The numbers follow what is called a *multinomial distribution*.

(C.ℓ) We do the same as in (C.2), but then consider the map

$$x \mapsto \ell x \bmod 1.$$

For riffle shuffles, we always have $\ell = 2^a$. But there is nothing special about powers of 2; the model covers the general case of an ℓ-shuffle. There are no mathematical difficulties with such a shuffle. But it is unclear whether there are any real world shuffles these models describe if there is no a, such that $\ell = 2^a$.

We are now almost able to write down the probability $Q_a(\pi_{\text{after}})$ for obtaining order π_{after} after a riffle shuffles (that is, after a 2^a-shuffle). Let us consider model (B.ℓ) with $\ell = 2^a$: We choose one out of $2^{52 \cdot a}$ sequences, each of them with the same probability. Thus,

$$Q_a(\pi_{\text{after}}) = N(\pi_{\text{after}})/2^{52 \cdot a},$$

where $N(\pi_{\text{after}})$ is the number of sequences that can yield the final order π_{after}. So how do we compute $N(\pi_{\text{after}})$? If π_{after} has the maximal number 2^a of rising sequences, the underlying sequence from (B.ℓ) is fully determined. For example, the deck order shown on page 38 allows us to reconstruct exactly which cards must have been in the top pile and which in the bottom pile. In that case, $N(\pi_{\text{after}}) = 1$. If there are fewer rising sequences than the maximal number 2^a, the sequence is not yet specified: There is at least one half-cut whose position is unclear. For example, if after a 2-shuffle, the deck of cards is still in its original order $1, 2, \ldots, 52$, the half-cut could have been at any position. Specifying this position, however, lets us identify the sequence from (B.ℓ). In general, $N(\pi_{\text{after}})$ is the number of possibilities to place $2^a - R(\pi_{\text{after}})$ cuts somewhere in the deck, where $R(\pi_{\text{after}})$ is the number of rising sequences of π_{after}. Or equivalently, it is the number of possibilities for placing the 52 cards in $52 + 2^a - R(\pi_{\text{after}})$ positions. This yields

$$Q_a(\pi_{\text{after}}) = \begin{cases} \binom{2^a + 52 - r}{52} \Big/ 2^{52 \cdot a}, & \text{if } \pi_{\text{after}} \text{ has } r \leq 2^a \text{ rising sequences} \\ 0, & \text{if } \pi_{\text{after}} \text{ has } r > 2^a \text{ rising sequences.} \end{cases} \tag{3.3}$$

3.6 MEASURING THE GOODNESS OF A SHUFFLE

As often happens, there is no universal definition of "goodness;" any useful definition must depend on the shuffle's purpose. In blackjack, for example, one could define the goodness of the shuffle as the difference between the expected gain under the optimal strategy for perfectly shuffled cards (–0.006) and the expected gain under the optimal strategy when one plays with a sorted deck on which a known type of shuffling has been performed. For example, for the half-shuffling described in equation (3.1), the player knows the full order of the deck of cards after seeing the first open card; the player therefore knows what the dealer's score will be and whether she should draw another card. In this sense, the half-shuffle is not "good," and it should certainly not be implemented by casinos.

For riffle shuffles, computing the expected returns of an optimal blackjack player is complicated. Even if we fully understand the shuffling and its resulting distributions, we need to know what the best strategy under the riffle-shuffling technique is. But this is unknown.

Instead, we define a shuffling scheme to be good if the resulting distribution is close to a uniform distribution. If this distance is zero (that is, the shuffling scheme is perfect), then the blackjack player cannot gain anything, so the above distance would be zero, too. Here we measure the distance between two distributions by the so-called total variation distance (TV). It is defined as

$$\delta(Q_a, Q_{\text{perfect}}) := \frac{1}{2} \sum_{\pi} \left| Q_a(\pi) - \frac{1}{52!} \right|.$$

In principle, we can now plug in formula (3.3) for Q_a and compute $\delta(Q_a, Q_{\text{perfect}})$. In practice, this is impossible, since the sum is taken over 52! permutations; even the fastest computers would not allow us to compute such a sum. But looking at equation (3.3), we realize: Most of the summands are the same! The probability $Q_a(\pi)$ depends only on the number of rising sequences of π. One can show that the number $A_{n,r}$ of permutations of n cards that have exactly r rising sequences is given by

$$A_{n,r} = \sum_{j=0}^{r-1} (-1)^j \binom{n+1}{j} (r-j)^n,$$

which is sometimes called the $(n, r-1)$th Eulerian number. This yields the TV distance

$$\delta(Q_a, Q_{\text{perfect}}) = \frac{1}{2} \sum_{r=1}^{52} A_{52,r} \left| \binom{2^a + 52 - r}{52} / 2^{52 \cdot a} - \frac{1}{52!} \right|,$$

a formula that is easily computable on a computer. The following figure shows the result:

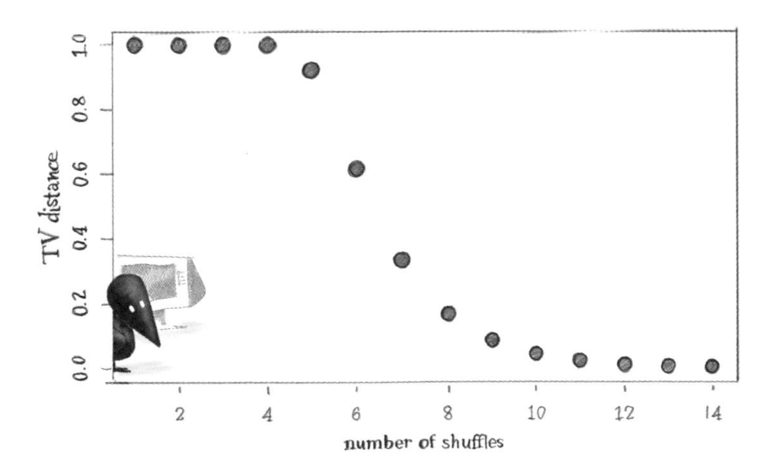

In terms of TV distance, four riffle shuffles are clearly insufficient. And while it is hard to set a precise boundary of when to call the TV distance small, different people have argued—based on the above figure—that 7 is a reasonable answer. Given the huge number of possible orders (52!, the number with more than 67 digits), 7 seems surprisingly small.

Here, we have mostly taken the sorted deck as the starting point before shuffling. All the above arguments, however, work exactly the same if the initial order of the cards is possibly different from random but is known. Thus, if you have a friend who is extremely good at memorizing cards (in particular, the order in which they have been played), make sure that you shuffle sufficiently often before starting a new game. This ensures that she does not have the chance to exploit the nonrandomness in the deck order. But then, with these abilities, your friend might be hard to beat in most card games, anyway.

In practice, there are many other shuffling techniques, too, of course. To the best of our knowledge, relatively little mathematical theory has been developed for these techniques.

3.7 SHORT HISTORY

The Gilbert-Shannon-Reeds model for riffle shuffles received its name from Edgar Gilbert, Claude Shannon, and Jim Reeds. Edgar Gilbert has written a monograph [Gilbert, 1955], and it is often stated (e.g., by Dave Bayer and Persi Diaconis; see below) that Jim Reeds developed the same model independently and wrote about it in an unpublished manuscript in 1981, sometimes referred to by the title "Theory of Riffle Shuffling." Unfortunately, we have not found that manuscript, so we cannot do anything other than restate this claim. The proof of the equivalence between the different descriptions of the model can be found in the beautiful work by Persi Diaconis, [e.g., Diaconis, 1988, lemma 2 and remark (f)]. The same work also contains data from 100 shuffles performed by Persi Diaconis and Jim Reeds that can be compared to the statistical model.

The first analysis of riffle shuffles in terms of total variation distance with respect to the uniform distribution is provided by Dave Bayer and Persi Diaconis [Bayer and Diaconis, 1992]. Often, David Aldous is also credited for this line of work [Aldous, 1983; Aldous and Diaconis, 1986]. As an alternative to the total variation distance, one can consider the Kullback-Leibler divergence. This measures the amount of information (in bits) about the initial order π_{before} that is contained in the shuffled deck π_{after}. In terms of the Kullback-Leibler divergence, riffle shuffles have been analyzed by Trefethen and Trefethen [2000] and by Stark et al. [2002]. The optimal strategy for blackjack was found by Baldwin et al. [1956].

3.8 PRACTICAL ADVICE

It takes a bit of practice to perform this trick reliably. In our experience, it helps if you go through the cards quickly without counting the precise score. If there are two (or more) cards that seem to be good candidates, the magician may want to spontaneously include this in her performance (e.g., by asking a discrimative question, such as "It is not a three, is it?"—"No."—"Certainly not, it is the jack of hearts!"). The trick also works, of course, if one does not start with a sorted deck, but with an order that one knows by heart. Then being able to spot the additional card becomes even more impressive.

4 ANIMAL STICKERS AND CYCLIC GROUPS

4.1 THE GAME

Number of players:	4–10
You will need:	1 animal sticker per player, each featuring a picture of an owl, tiger, or zebra; 3 boxes, lined up on the ground

The audience attaches 1 animal sticker to the back of each player, so that no player can see his or her own animal. In principle, every player could have the same animal sticker. As in previous chapters, the players can only win if they work as a team. The players are free to walk around the room and look at the stickers of other players but are not allowed to look at their own sticker. Three boxes have been placed in a line on the ground. After 1 minute, each player is asked to line up behind 1 of the 3 boxes. The group wins if and only if the players in each of the 3 groups all have the same animal. For example, if all players behind box 1 have a tiger on their back, all players behind box 2 have a zebra, and no one stands behind box 3, the game is won. But if behind one of the boxes there is both a player with a zebra and a player with an owl, the game is lost.

What is the best strategy for the group? If the players use that strategy, how often will they win?

In general, there will be p animals, p boxes, and n players. But first, let us look at the example above, where we have $p=3$ animals (zebra, owl, and tiger) and $n=4$ players: Anwar, Bella, Charlie, and Deepti. We assume first that the audience member is choosing the animal for each player randomly from the set of 3 animals. A possible strategy for the players would be as follows: they prearrange that Anwar and Bella are going to line up behind the first box, Charlie will take the second and Deepti the third. What is their chance of success with this strategy? Suppose that the audience member chose a zebra for Anwar. Then Bella also needs to have a zebra for the group to win, and neither Charlie nor Deepti can have a zebra. Also, the animals for Charlie and Deepti need to be different. This leaves two options for the animals of Bella, Charlie, and Deepti: either zebra, owl, tiger or zebra, tiger, owl. Both of these sequences have a probability of $1/3 \cdot 1/3 \cdot 1/3 = 1/3^3 = 1/27$ if the animals are chosen independently for each player, so that the overall chance of winning is $2/27$, which is below 8%. If Anwar gets a different animal, the same argument applies, so the overall chance of winning is $2/27$.

Thus, lining up behind the boxes in prearranged configurations will not be a good strategy, and the probability of winning will fall further for more players.

Choosing a different configuration will not help either. For example, if all 4 players line up behind the same box, the chance is only 1/27, as Bella, Charlie, and Deepti then all need to have the same animal as Anwar, whereas spreading out across more boxes leaves more options for success.

The key to success thus is to exploit what players are observing about the animals of other players. At first, this seems unlikely to be helpful, as the animals of Anwar, Charlie, and Deepti do not hold any information about the animal of Bella if animals are chosen randomly by the audience. So what can Bella learn by her observations? Remember that Bella does not have to guess her animal in this game, she just has to line up behind the same box as other players with her animal. So what does Bella have in common with the players who have the same animal? What about the general case, where there are p animals, p boxes, and n players? We encourage the reader to pause here and think a bit more about this game.

4.2 SOLUTION FOR 3 ANIMALS

Let us revisit the example including Anwar, Bella, Charlie, and Deepti. The key insight is the following. If Bella observes 2 zebras and 1 tiger on the other players, say, then she will know that any other player with the same animal as hers (if there are any) will also observe 2 zebras and 1 tiger, irrespective of which animal is hers. And players with a different animal will see a different configuration of animals. For example, if she had a zebra on her back, and so did Anwar, then Anwar would also see 1 tiger and 2 zebras (instead of his own zebra, he will see Bella's). If she has an owl on her back, then nobody else will see 2 zebras and a tiger, as all of them will see her owl among the 3 animals they observe. The players thus need to be able to translate the collection of animals they observe into a choice of the box, so that players seeing the same collection of animals pick the same box, and players seeing a different collection of animals do not pick the same box. This can be done using numbers and so-called "modulo" computations.

First, we can attribute to each animal a value in $\{0, 1, 2\}$:

zebra is Zero, owl is One, tiger is Two.

Taking the previous example, the situation is as follows:

$$\text{Anwar has a zebra} \equiv 0,$$

$$\text{Bella has an owl} \equiv 1,$$

$$\text{Charlie has a tiger} \equiv 2,$$

$$\text{Deepti has a zebra} \equiv 0.$$

Having seen the animals on the other players, every player lines up behind 1 of the 3 boxes. To do so, they pretend that the boxes are numbered 0, 1, and 2 (left to right from the players viewpoint). Anwar can see that Bella has an owl on her back, Charlie has a tiger, and Deepti has a zebra. His choice of which box to stand behind now depends on the animals he sees on the other players. But how can Anwar choose a box so that everybody else with a zebra (Deepti in the example) will choose the same box as him, and all the players with a different animal will choose a different box?

As mentioned above, the key is that players with the same animal will see the same collection of animals. As a strategy, the players can, for example, sum up the numeric value of all the animals they are seeing. For Anwar, this means

$$1\,(\text{Bella's owl}) + 2\,(\text{Charlie's tiger}) + 0\,(\text{Deepti's zebra}) = 3.$$

The computations performed by the 4 players are summarized in the following table:

	Anwar	Bella	Charlie	Deepti	Sum
Anwar can see	✗	1 (owl)	2 (tiger)	0 (zebra)	3
Bella can see	0 (zebra)	✗	2 (tiger)	0 (zebra)	2
Charlie can see	0 (zebra)	1 (owl)	✗	0 (zebra)	1
Deepti can see	0 (zebra)	1 (owl)	2 (tiger)	✗	3

Anwar and Deepti will both arrive at a numeric value of 3. They now have to translate this numeric value into a choice of 1 of the 3 boxes, making sure that players who see a different collection of animals (different by, at most, only 1 animal) will choose a different box. Here, the observed sums could range between value 0 (only seeing zebras on all other players) to 6 (all tigers). So, how can these numbers be distributed among 3 boxes?

Let us say that 2 numbers are equivalent if they are equal modulo 3, which means that their difference is a multiple of 3. The numbers −1, 2, 5, 8, 14, and 602 are therefore equivalent, for example, and are said to be in the same equivalence class. Looking only at nonnegative integer values, the equivalence classes in our example are

$$[0] = \{0, 3, 6, \dots\},$$

$$[1] = \{1, 4, 7, \dots\},$$

$$[2] = \{2, 5, 8, \dots\}.$$

Section 4.3 contains a further, more precise definition of equivalence classes. For the solution, we now associate box 0 with equivalence class [0], box 1 with equivalence class [1], and box 2 with equivalence class [2]. Players have to line up behind the box whose equivalence class contains the sum of the animal values they are seeing on all other players, excluding themselves.

In the example, Anwar translates the sum of 3 into equivalence class [0] and lines up behind box 0. Bella obtains a sum of 2 and hence lines up behind box 2. Charlie will see a sum of 1 and lines up behind box 1, and Deepti lines up behind box 0 (since her sum equals 3). All players are grouped correctly.

This is not a coincidence. If they use this strategy, the team will win every time, no matter how the animal stickers are distributed. To prove this, we need to show that (a) players with different animals choose different boxes, and (b) players with the same animals choose the same box. The second part is easy to see: players with the same animal will see the same collection of other animals and therefore will arrive at the same sum and hence the same box.

So why do players with different animals always choose different boxes? What if they arrive at different sums but the boxes associated with these sums are identical? To see that this cannot happen, consider the difference between the sums of 2 players. It is largest when one carries an animal with a value of 0 and the other with a value of 2. The difference between the two sums is then 2 at most. Two different numbers are only equivalent, however, if they differ by a multiple of 3. Thus, the two players cannot arrive at the same box if they carry different animals.

For the more general case, we have p boxes and p animals and equivalence classes

$$\text{box } 0 \text{ associated with } [0] = \{0, p, 2p, \ldots\},$$

$$\text{box } 1 \text{ associated with } [1] = \{1, p+1, 2p+1, \ldots\},$$

$$\vdots$$

$$\text{box } p-1 \text{ associated with } [p-1] = \{p-1, 2p-1, 3p-1, \ldots\},$$

and, as we discuss below, all the above arguments and strategies still apply.

We will now introduce some mathematics that will help us solve the general case of p animals and n players. If you are instead interested in a different variation of the game, feel free to skip the following section and jump directly to section 4.4.

4.3 SOME MATHEMATICS: CYCLIC GROUPS

Equivalence Classes

In the example above, we did not distinguish between the sums 1, 4, or 7, because all of them implied that the players lined up in front of the second box. Mathematically, we have created an equivalence relation on the integers $\mathbb{Z}=\{\ldots,-2,-1,0,1,2,\ldots\}$. We say that $\sim$ is an *equivalence relation* on $\mathbb{Z}$ if it satisfies the following three properties:

1. *Reflexivity*: $a \sim a$ for all $a \in \mathbb{Z}$.
2. *Symmetry*: if $a \sim b$, then also $b \sim a$ for all $a, b \in \mathbb{Z}$.
3. *Transitivity*: if $a \sim b$ and $b \sim c$, then also $a \sim c$ for all $a, b, c \in \mathbb{Z}$.

These three properties are clearly fulfilled if we say two numbers a and b are equivalent if their difference $a-b$ is divisible by 3. Instead of writing $3 \sim 0$ and $11 \sim 2$, it is common to write

$$3 \equiv 0 \bmod 3, \text{ and } 11 \equiv 2 \bmod 3,$$

which is also called taking numbers *modulo* 3. All integers that are equivalent to a form the equivalence class

$$[a] := \{b \in \mathbb{Z} : b \sim a\}.$$

In our example, the equivalence class of [0] is $[0]=\{\ldots,-3,0,3,6,\ldots\}$, and the equivalence classes of 2 and 11 are identical: $[11]=[2]$. It is possible to perform calculations with equivalence classes, just like we can calculate with numbers: they form a group.

Groups

A set G and an operation $\oplus$ form a *group* if the following holds:

1. *Closure*: for all $x, y \in G$, we have $x \oplus y \in G$.
2. *Identity*: there exists an identity element $1 \in G$ such that for all $x \in G$, it holds $x \oplus 1 = 1 \oplus x = 1$.

3. *Associativity*: for all $x, y, z \in G$, we have $(x \oplus y) \oplus z = x \oplus (y \oplus z)$.
4. *Inverse*: for all $x \in G$, there exists an element x^{-1} with $x \oplus x^{-1} = x^{-1} \oplus x = 1$.

For our example, the relevant set G is the set of all distinct equivalence classes if using the modulo 3 equivalence relation. These equivalence classes are [0], [1], and [2]. The relation $\oplus$ in this case is defined by

$$[a] \oplus [b] := [a+b], \tag{4.1}$$

for example, $[2] \oplus [2] = [2+2] = [4] = [1]$. All four properties are then satisfied.[1] The identity element is [0], and the inverse of $[a]$ is $[-a]$. Our group G is often denoted by $\mathbb{Z}/3\mathbb{Z}$ or $\mathbb{Z}_3$, and is usually referred to as "$\mathbb{Z}$ modulo $3\mathbb{Z}$".

Cyclic Groups

A *cyclic group* G is a group that can be generated by a single element. In our example, all members of the set can be generated by the element [1], as $[0] = [3] = [1] \oplus [1] \oplus [1]$ and $[2] = [1] \oplus [1]$.

Revisiting the Solution

We are now able to rewrite the solution above with the new notation. First, let us denote the animals of Anwar, Bella, Charlie and Deepti by

$$C_{Anwar}, C_{Bella}, C_{Charlie}, \text{ and } C_{Deepti},$$

respectively. Then, each player calculates the sum of the other players' animals as follows:

$$\begin{aligned} S_{Anwar} &= C_{Bella} + C_{Charlie} + C_{Deepti} \\ &= 1 \text{ (owl)} + 2 \text{ (tiger)} + 0 \text{ (zebra)} = 3, \end{aligned}$$

---

1. Strictly speaking, we would also need to show that our operation is well defined: if $[a] = [\tilde{a}]$ and $[b] = [\tilde{b}]$, then $[a+b] = [\tilde{a}+\tilde{b}]$. Otherwise, our definition (4.1) would not make any sense. Please have a go at proving this statement.

$$\begin{aligned}
S_{Bella} &= C_{Anwar} && && + C_{Charlie} && + C_{Deepti} && \\
&= 0 \text{ (zebra)} && && + 2 \text{ (tiger)} && + 0 \text{ (zebra)} && = 2, \\
S_{Charlie} &= C_{Anwar} && + C_{Bella} && && + C_{Deepti} && \\
&= 0 \text{ (zebra)} && + 1 \text{ (owl)} && && + 0 \text{ (zebra)} && = 1, \\
S_{Deepti} &= C_{Anwar} && + C_{Bella} && + C_{Charlie} && && \\
&= 0 \text{ (zebra)} && + 1 \text{ (owl)} && + 2 \text{ (tiger)} && && = 3.
\end{aligned}$$

Player k then chooses a box $b \in \{0, 1, 2\}$ such that $[b] = [S_k]$. So, $b = 0$ for Deepti, for example, since $[0] = [S_{Deepti}] = [3]$.

Solution to the General Case

Somewhat surprisingly, the game still works for p animals, p boxes, and n players, no matter how large p and n are. The solution can then be formulated as follows. Let $C_k \in \{0, \ldots, p-1\}$ be the value of the animal on player k. Let further

$$S_k := \sum_{\ell=1;\ell \neq k}^{n} C_\ell = \sum_{\ell=1}^{n} C_\ell - C_k$$

be the sum of the animal values on all players other than player k. The box $b \in \{0, \ldots, p-1\}$ chosen by player k is the box for which the equivalence classes agree, that is $[b] = [S_k]$. Instead of $\mathbb{Z}/3\mathbb{Z}$, we here consider the group $\mathbb{Z}/p\mathbb{Z}$, that is, two numbers are equivalent if their difference can be divided by p. If $p = 7$, for example, and a player computes the sum 18, she should then choose box number 4, since

$$[4] = [18]$$

(because $18 - 4$ can be divided by 7). We then have the equivalences

$$[S_k] = [S_{k'}] \quad \Leftrightarrow \quad S_k = S_{k'} \quad \Leftrightarrow \quad C_k = C_{k'},$$

so that the equivalence classes of two players match if and only if they have the same animal on their backs. $[S_k] = [S_{k'}]$ implies $S_k = S_{k'}$, because the values of two different sums S_k and $S_{k'}$ can differ by at most $p - 1$. Two players will hence choose the same box if and only if their animals match.

4.4 VARIATION: COLORED HATS IN A LINE

Number of players:	4–10
You will need:	colored hats in 3 (or more) unique colors

All n players form a line and members of the audience choose a colored hat for each player; several players may receive the same color. There are p distinct hat colors, and everybody knows what the color options are.

As illustrated above, each player can see the color of the hats in front of them. They can neither see the color of their own hat nor the colors of any of the other players behind them. The person at the front of the line cannot see any of the hats. Each player guesses their color, starting with the person at the back. All the players can hear all the answers. The group can agree together on a strategy beforehand to make sure that as many people as possible will guess their color correctly.

What is the best strategy, and how many colors can the group guess correctly? Is there a way to make sure that a certain number of colors are always guessed correctly every time the game is played?

How Well Can a Strategy Work?

The general idea is similar to the animal sticker game. However, it is clear that, generally speaking, not all players will be able to guess their colors correctly. The player at the back of the line is in a tricky situation. She has to make her guess first and can base her decision only on the colors of the hats in front of her. If we assume that the colors in front of her are chosen independently of her own hat color, then the observed colors will not contain any information about her own color. Even worse, if the audience guesses the team's strategy, they can make the player in the back answer incorrectly all the time. If the player forgets about the group and thinks only about herself, all she can do is guess randomly. In this case, her choice will be wrong with probability $1 - 1/p$ for p distinct colors. We thus have a bound on the expected number[2] of mistakes:

$$\mathbf{E}(\text{mistakes}) \geq 1 - \frac{1}{p}.$$

As the number of correct guesses is equal to n minus the number of mistakes, we get an upper bound on the number of correct guesses:

$$\mathbf{E}(\text{correct guesses}) \leq n - 1 + \frac{1}{p}.$$

Can we find a strategy that achieves this bound in that there will be on average $n - 1 + 1/p$ correct guesses among all n players?

Solution

Let us start by discussing the solution for the case where three hat colors (zucchini, orange, and turquoise) are chosen randomly for four players (Anwar, Bella, Charlie, and Deepti). We first assign a numerical value to a color in the following way:

2. See also "What Is ...an Expectation?" (appendix B.6).

zucchini is Zero, orange is One, and turquoise is Two,

just as we did in the animal sticker game. If you have not heard of the color "zucchini" before, you can imagine it as being green (it is unfortunately the only color we could think of that starts with a "z"). The distribution of hats is:

Anwar	Bella	Charlie	Deepti
zucchini	orange	turquoise	zucchini

Every player now has to guess their color, starting with Deepti, the player at the back.

The argument above implies that the first player to guess her color, Deepti, has no better chance to get her color right than if she were randomly guessing her color. One strategy for her would indeed be to just guess a color at random and hope that the guess is correct. There would be a high probability that she would be wrong, namely $1 - 1/p$. Even worse, Charlie in front of her then again has no information about his own color and is in the same clueless state that Deepti is in. If Charlie and subsequent players then repeat the random guessing strategy, we can hardly expect many correct guesses.

So far, we have not taken advantage of the fact that there is a key difference between Charlie and Deepti. Charlie can base his guess on both the colors of Anwar's and Bella's hats in front of him and also on the guess of Deepti. The colors of Anwar's and Bella's hats in front of him were chosen independently of Charlie's color and therefore contain no information about the color of Charlie's when considered on their own. Therefore, the key has to be to create a dependence between Deepti's guess and Charlie's hat. The dependence is only relevant if Deepti's guess is based on the color of Charlie's hat.

How can we get Deepti to make a guess that will help Charlie work out the color of his hat? One possibility is that Deepti just announces the color of Charlie's hat in front of him instead of trying to guess his own color. Then Charlie can guess his own color correctly, as he just has to repeat the color that Deepti just said. However, a downside of this strategy is then that Bella again has no information about the color of her hat. She will have to repeat the procedure by announcing the color of Anwar's hat in front of her, which Anwar can then repeat and guess his color correctly. Following this strategy, we guarantee that at least half of the guesses will be correct (Anwar's and Charlie's). Bella's and Deepti's guesses will be correct only if by chance their hat color matches Anwar's and Charlie's color, respectively.

It turns out that we can can do better than that and, in fact, we can match the previous bound by making sure that everybody except possibly Deepti can guess their color correctly!

To avoid Bella having to start from scratch, we clearly have to use a different strategy, and we can use one similar to that used in the animal sticker game. The difference is now that the players only have partial information. Consider the table of colors that the players can see:

	Anwar	Bella	Charlie	Deepti	Sum
Deepti can see	0	1	2	✗	$3 = S_{Deepti}$
Charlie can see	0	1	✗	✗	$1 = S_{Charlie}$
Bella can see	0	✗	✗	✗	$0 = S_{Bella}$
Anwar can see	✗	✗	✗	✗	$0 = S_{Anwar}$

Here and below, we denote the sum of all observed colors for player k by S_k and the player k's true color by C_k. The associated colors are again the equivalence classes

$$\text{zucchini color: } [0] = \{\dots, -3, 0, 3, 6, 9, 12, \dots, \},$$

$$\text{orange color: } [1] = \{\dots, -2, 1, 4, 7, 10, 13, \dots, \},$$

$$\text{turquoise color: } [2] = \{\dots, -1, 2, 5, 8, 11, 14, \dots\}.$$

The first player to announce, Deepti, now announces the color of her observed sum

$$\text{Deepti announces: } [S_{Deepti}] = [3] = [0] = \text{zucchini}.$$

Perhaps surprisingly, the next players can now calculate their colors. First, the difference between S_{Deepti} and $S_{Charlie}$ is all down to the color of Charlie's hat. Thus

$$\text{Charlie announces: } [S_{Deepti} - S_{Charlie}] = [0 - 1] = [2] = \text{turquoise},$$

and he gets his color correctly. Here, it does not matter that he substitutes 0 for the color zucchini that Deepti announced instead of the value 3 that Deepti had in mind, because both belong to the same equivalence class. In fact, he could take any member of the equivalence class of color zucchini instead of the value 0 and still arrive at the correct solution.

Continuing like this, Bella can announce her color correctly as

$$\begin{aligned} \text{Bella announces: } & [S_{Deepti} - C_{Charlie} - S_{Bella}] \\ & = [0 - 2 - 0] = [-2] = [0] = \text{ orange}. \end{aligned}$$

Note that she has to subtract the guessed color of Charlie, not the sum Charlie has been seeing. Why is the guess correct for Bella? The sum that Deepti sees is

$$S_{Deepti} = C_{Charlie} + C_{Bella} + C_{Anwar} = C_{Charlie} + C_{Bella} + S_{Bella},$$

and thus, substracting $C_{Charlie}$ and S_{Bella} on both sides,

$$C_{Bella} = S_{Deepti} - C_{Charlie} - S_{Bella}.$$

The equivalence class (color) of C_{Bella} is therefore equal to the equivalence class of the right-hand side, where we replace the correct values, S_{Deepti} and $C_{Charlie}$, by any arbitrarily chosen member of their respective equivalence classes. The color $[S_{Deepti}]$ is announced by Deepti, and the color $[C_{Charlie}]$ is announced by Charlie. To arrive at her own color, Bella is thus effectively subtracting the color that Charlie announced and the sum of colors she has seen from the color that Deepti announced first.

Finally, Anwar can guess his color correctly as

$$\text{Anwar announces: } [S_{Deepti} - C_{Charlie} - C_{Bella} - S_{Anwar}]$$
$$= [0 - 2 - 1 - 0] = [-3] = [0] = \text{zucchini}.$$

Anwar has to subtract the colors $[C_{Charlie}]$ and $[C_{Bella}]$ that Charlie and Bella announced and the sum of colors he has seen from the color $[S_{Deepti}]$ announced by Deepti.

The scheme can also be used with a general number $p \in \mathbb{N}$ of different colors and $n \in \mathbb{N}$ players. The sums of the observed colors of player k is defined as above:

$$S_k := C_{k-1} + C_{k-2} + \cdots + C_1.$$

Player n in the back announces the equivalence class of S_n, and subsequently, player k announces the equivalence class of

$$S_n - C_{n-1} - C_{n-2} - \cdots - C_{k+1} - S_k.$$

Because

$$S_n = C_{n-1} + C_{n-2} + \cdots + C_{k+1} + C_k + S_k,$$

this will always be the correct color.

All players will thus announce their color correctly except, possibly, for the player at the back of the line. This guarantees that out of n players, at least $n-1$ players will guess their color correctly! The player in the back can be right, too, if his color matches the color of the sum of all colors in front of him (that is, if $[S_n] = [C_n]$). If the color of the hats are chosen independently, this will happen with probability $1/p$. The expected number of successes with this scheme is based on the assumption of independently chosen colors $n - 1 + 1/p$, which matches the upper bound we derived in section 4.4. Thus, the scheme cannot be improved.

A (not very practical but mathematically interesting) version for infinitely many players can be found in appendix C.2.

Adversarial Audience

If the audience expects the team to use the above strategy, it can of course provide a color to the player in the back that is not in the equivalence class of S_n. This way, the last player will certainly answer incorrectly, and the expected number of successes is down to $n-1$. There is, however, a way out for the team: The players first agree on a random number R between 0 and $m-1$. Player n in the back then announces S_n+R instead of S_n, and all players adjust accordingly. This way, the expected number of successes is back to $n-1+1/p$, independently of how the audience distributes the colors.

4.5 SHORT HISTORY

The origin of the hats-in-a-line game could not clearly be traced, but both games are mentioned, for example, in two papers in the *Electronic Journal of Combinatorics*, namely, Aravamuthan and Lodha [2006] and Paterson and Stinson [2010]. The latter also introduce a new game that sits somewhere between the hats-in-a-line game and the hat-color game of section 4.1. The idea for solving the infinite hat game presented in section C.2 can be found in the article by Hardin and Taylor [2008], which also contains further references on infinite hat problems.

4.6 PRACTICAL ADVICE

We can, of course, use colored hats instead of animal stickers for the animal sticker game (and vice versa).

Variations on the Animal Sticker Game

In the version described, the players line up behind the boxes jointly and are able to see where everybody else is lining up. In a seemingly more difficult version, the players sit down and then close their eyes (or the other way around, as they like) and are called one after another to choose a box, without being able to see or hear the choices of other players. They can, of course, use the same strategy here, and the game is thus not more difficult to play.

An additional variation is as follows: The first player to choose a box can look at her animal after having chosen a box. She then has to mark all three boxes with a zebra, tiger, or owl, so that only the audience but not the players can see the markings. The remaining players then have to line up behind the boxes that are marked with their animal.

She is, of course, marking the box she chose with her animal and the others in reverse order of their numeric values. In other words, after having put her own animal on her chosen box, the box with a numeric value 1 *lower* than her own box (first box if she had the second, second box if her own was the third, third box if her own box was the first) is marked with the the animal with a numeric value 1 *higher* (so an owl if her own animal is the zebra, a tiger if her own animal is an owl, and a zebra if her own animal is a tiger). Analogously, then the box with a numeric value *lower* by 2 (or equivalently, higher by 1) is paired with the animal with numerical value *higher* by 2 (or equivalently, lower by 1). The remaining players will play just as in the original game and will end up behind the box with the correct label.

All of the above versions assume that the boxes are either numbered 0, 1, and 2, or have a particular order (from left to right, say). There is a variation that works without having such an order. If any animal can be assumed to appear at least once, one can perform a slight variation of the game. One player with a 0 can take a look at her sticker (cheating) and say: "I have got a …" (announcing the animal he finds). Then all players with a zero could follow: "we do, too!" Then the group of players with a 1 follows, and so on. That is, each player can announce her own animal—without having to line up behind any box.

Variation on the Colored Hat Game

The first player to announce (at the back of the line) is generally wrong (unless he is lucky), while all other players get their colors right. The first player to announce can, of course, say something along the lines of "I would have guessed 'turquoise' but I am quite sure it is wrong," in this way signaling the color turquoise to the other players while acknowledging that he is probably wrong.

Faster Modulo Operation in Both Games

Taking the modulo operation can take some time if the sums involved grow very large. This can be problematic for a large number of players. It becomes easier to play the games if we also use negative values, such as replacing

$$\text{zebra} \equiv 0, \quad \text{owl} \equiv 1, \quad \text{tiger} \equiv 2$$

with

$$\text{zebra} \equiv 0, \quad \text{owl} \equiv 1, \quad \text{tiger} \equiv -1.$$

or, to make it memorable,

$$\text{zebra} \equiv 0 \text{ (zero)}, \quad \text{owl} \equiv 1 \text{ (one)}, \quad \text{mole} \equiv -1 \text{ (minus one)}.$$

Then the players only have to count how many more owls than moles they can see (and take the result modulo 3).

5 OPERA SINGERS AND INFORMATION THEORY

5.1 THE GAME

Number of players:	5–20, plus 1 game show host with complete knowledge of what is going on.
You will need:	a stage divided into 2 separate areas, left and right, and 1 off-stage area (backstage or frontstage)

The players represent a group of opera singers. Among them are two stars. These stars are either friends or enemies. Only the host knows who the stars are and the nature of their relationship (friends or enemies). So in this game, not even the two stars know each other: The stars do not know that they are stars themselves, they do not know who the other star is, and they do not know their relationship. Neither do the other singers know. The singers' common goal is to identify who the stars are and whether they are friends or enemies. To do so, they send two groups of singers to the different areas on the stage. The group on the left represents the singers from the first half of the opera, the group on the right from the second half. The remaining players in the backstage area do not take part in that opera performance. The host has complete knowledge of who's who and now reveals some information on the players' identities. He does so by providing feedback on each of these "performances" by classifying each performance as a "big success," "disaster," or "neutral," according to the following two rules:

1. When no stars or only one star appears on stage, the critics will not even write about the opera, and the host will classify the performance as "neutral."
2. If both stars appear on stage, the critics will be excited, and the opera performance will not be considered neutral. If the stars are friends and appear in the same half or if the stars are enemies and appear in different halves, the critics will declare the opera to be a "big success." Otherwise (that is, if the friends sing in different halves or the enemies sing together), the opera will be deemed a "disaster."

Let us consider two examples: In the first example, there are two enemies (called E1 and E2), but only one of them is on stage, as shown in the figure below on the left. Here, the outcome is neutral. In the second example, there are two friends (F1 and F2) singing in the same half, as shown on the right. This opera is a big success.

Outcome: neutral

Outcome: big success

After receiving the critique, the group decides whom to send on stage next. The group of players is asked to develop a strategy that allows them to find out as quickly as possible who the stars are and whether they are friends or enemies. Fortunately, so many operas with two halves have been composed that there are no constraints on the number of singers in each group: The numbers of singers can differ between the two halves (as in the example above) and between performances. One or two of the groups of singers may even be empty.

What is the optimal strategy for 5 singers? More precisely, which strategy requires the lowest number of performances, even in the worst case? Is there an easy strategy that comes close to optimal? What about the optimal strategy for 100 singers?

For simplicity, let us first look at the example with 5 singers. It does not even seem obvious how to find a strategy that can be described easily. We can, for example, select a pair of singers, schedule them for the first half of the opera, and leave the second half empty. We can then repeat this strategy for all pairs of singers. In most cases, at least one of the stars will be left outside, and the outcome of the performance will be neutral. If, however, the outcome is not neutral, we immediately know that this pair of singers must be the pair of stars we are looking for, and we also know whether they are friends (the opera is a success) or enemies (it is a disaster). In the worst case, we have a nonneutral outcome only for the last pair of singers. Given that there are 5 singers in total, how many pairs of singers are there? There are 5 possibilities to choose the first singer of the pair and 4 possibilities to choose the second singer from the remaining singers, which gives $5 \cdot 4 = 20$ pairs. This, however, counts every pair exactly twice, since for forming a pair, it does not matter which singer was chosen first and which one second. In total, there are thus $5 \cdot 4/2 = 10$ pairs. This number is often denoted by $\binom{5}{2} = 10$ (see appendix B.4: What Is ... a Binomial Coefficient?).

Thus, with this strategy, we obtain the correct answer after at most 10 performances.

This strategy, however, is not optimal. One of the explanations for this is that each performance carries only a little "information" on average (we will make this term precise later), since most of the performances will be considered neutral. One possibility to improve on the above strategy is to look at 2 pairs at the same time; that is, to have one pair in the first half and another in the second half. After 5 performances, we have seen all singers on stage and have, even in the worst case, identified a group of 4 singers that must contain the 2 stars. Using 2 more performances, it is possible to identify the 2 stars and their relation. This strategy thus works in at most $5+2$ attempts. We will see that this strategy is not optimal either.

We encourage the reader to pause here. Using the argument above, 7 is an upper bound on the number of performances required in the worst case. Is there a way to develop a lower bound?

5.2 HOW WELL CAN A STRATEGY WORK?

We first compute a lower bound on what we can hope for: What is the minimum number of performances required for us to be certain about who the stars are and what their relationship is? Let us assume that there are 5 singers. Then we can count the number of possible solutions. There are $5 \cdot 4/2 = 10$ different pairs that could be the pair of stars. Each pair of stars can either consist of 2 friends or 2 enemies, so there are 20 possible solutions in total.

Now, each opera performance can be seen as an experiment with 3 possible outcomes (neutral, disaster, or success). Suppose there is a deterministic strategy that will always enable us to know the true solution after 2 performances. Then, each of the 20 possible solutions will yield exactly 1 pair of outcomes, such as (success, neutral), (success, disaster), or (neutral, neutral). There are, however, only $3 \cdot 3 = 9$ possible outcomes with 2 performances. Several solutions must yield the same pair of outcomes, and we have no chance of distinguishing these solutions. As a consequence, there is no strategy that always succeeds with just 2 performances.[1] What about 3 performances? We then have $3 \cdot 3 \cdot 3 = 27$ possible outcomes, which could be enough to discover the truth among the 20 possibilities. We will see that there is indeed a strategy that will always find the truth after 3 performances. Section 5.3 presents the solution in the case of 5 singers and 3 performances. In section 5.4, we introduce some ideas from information theory that will help us understand why these strategies work and how to extend them to more general settings.

5.3 SOLUTION FOR 5 SINGERS

First, let us write down all possible pairs of opera singers together with an indicator (+ or −) as to whether they are friends (+) or enemies (−). We stated already that there are 20 hypotheses in total: $\Omega := \{(1,2,+), (1,2,-), (1,3,+), (1,3,-), (1,4,+), (1,4,-), (1,5,+), (1,5,-), (2,3,+), (2,3,-), (2,4,+), (2,4,-), (2,5,+), (2,5,-), (3,4,+), (3,4,-), (3,5,+), (3,5,-), (4,5,+), (4,5,-)\}$. A strategy that lets us identify the true solution in any of these cases is summarized in the following figure.

1. The same argument holds for randomized strategies. Each solution then yields a probability distribution over the 9 pairs of outcomes. Therefore, there will be at least 1 pair of outcomes that has a positive probability of occurring for at least 2 different solutions. For simplicity, however, we are going to restrict ourselves to deterministic strategies in the remainder of this chapter.

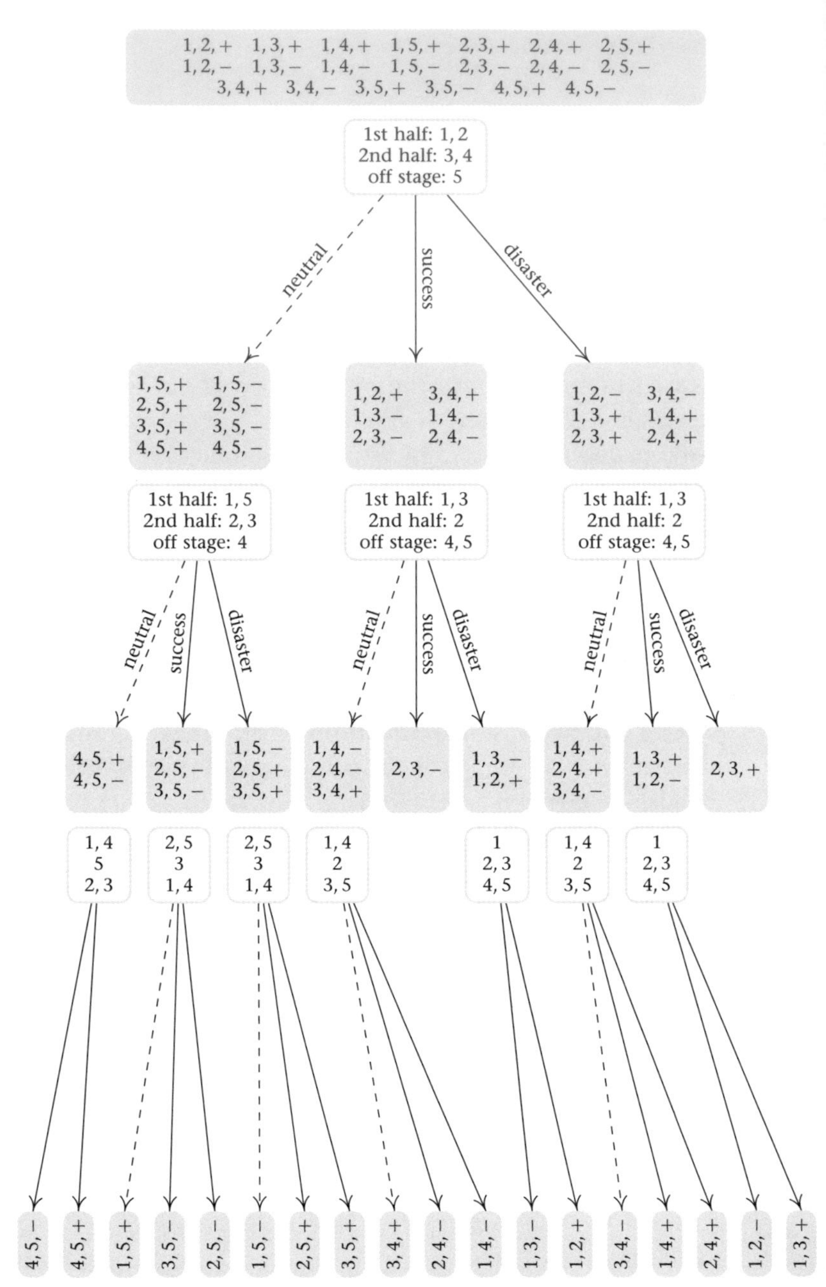

1, 2, + 1, 3, + 1, 4, + 1, 5, + 2, 3, + 2, 4, + 2, 5, +
1, 2, − 1, 3, − 1, 4, − 1, 5, − 2, 3, − 2, 4, − 2, 5, −
3, 4, + 3, 4, − 3, 5, + 3, 5, − 4, 5, + 4, 5, −
1st half: 1, 2
2nd half: 3, 4
off stage: 5
neutral
success
disaster
1, 5, + 1, 5, −
2, 5, + 2, 5, −
3, 5, + 3, 5, −
4, 5, + 4, 5, −
1, 2, + 3, 4, +
1, 3, − 1, 4, −
2, 3, − 2, 4, −
1, 2, − 3, 4, −
1, 3, + 1, 4, +
2, 3, + 2, 4, +
1st half: 1, 5
2nd half: 2, 3
off stage: 4
1st half: 1, 3
2nd half: 2
off stage: 4, 5
1st half: 1, 3
2nd half: 2
off stage: 4, 5
neutral
success
disaster
neutral
success
disaster
neutral
success
disaster
4, 5, +
4, 5, −
1, 5, +
2, 5, −
3, 5, −
1, 5, −
2, 5, +
3, 5, +
1, 4, −
2, 4, −
3, 4, +
2, 3, −
1, 3, −
1, 2, +
1, 4, +
2, 4, +
3, 4, −
1, 3, +
1, 2, −
2, 3, +
1, 4
5
2, 3
2, 5
3
1, 4
2, 5
3
1, 4
1, 4
2
3, 5
1
2, 3
4, 5
1, 4
2
3, 5
1
2, 3
4, 5
4, 5, −
4, 5, +
1, 5, +
3, 5, −
2, 5, −
1, 5, −
2, 5, +
3, 5, +
3, 4, +
2, 4, −
1, 4, −
1, 3, −
1, 2, +
3, 4, −
1, 4, +
2, 4, +
1, 2, −
1, 3, +

We start by putting singers 1 and 2 in the first half, and singers 3 and 4 in the second half (see the white box near the top of the figure). In the scenario where the opera is declared a success, for example, we are left with 6 possible hypotheses (see the gray box connected by "success"), and we then put singers 1 and 3 in the first half and singer 2 in the second half (see the white box below). No matter what the underlying truth is, after no more than 3 performances, there is only one hypothesis left, and so the group will know exactly who the stars are and whether they are friends or enemies.

5.4 SOME MATHEMATICS: INFORMATION THEORY

The key to solving this game is to ask the right questions. Similarly, in the game 'Who am I?', players receive hidden identities (e.g., based on famous people) and try to reveal their identities by asking questions that will provide as much information as possible. But how do we measure informativeness? This is the starting point of *information theory*.

Measuring Information Content

Suppose you are playing a number-guessing game with your friend. Your friend is thinking of a number between 0 and 15, and your task is to find which number your friend is thinking of by asking yes/no questions. The goal is to ask as few questions as possible. Let us assume that your friend chooses her number at random and that she does not prefer any of the numbers: None of the 16 numbers is chosen with larger probability than the others.

To help us understand information content, it will be convenient to write your friend's number as a *binary word*. If she thinks about the number 5, we write this as 0101. Why 0101? Because 5 can be written as

$$5 = \mathbf{0} \cdot 2^3 + \mathbf{1} \cdot 2^2 + \mathbf{0} \cdot 2^1 + \mathbf{1} \cdot 2^0.$$

Similarly, 14 is written as 1110, since

$$14 = \mathbf{1} \cdot 2^3 + \mathbf{1} \cdot 2^2 + \mathbf{1} \cdot 2^1 + \mathbf{0} \cdot 2^0.$$

The four digits in 0101 and 1110 are often called "bits," which is short for "binary digits." (You can find more details about binary expansions in "What Is ... a Binary Number?" in appendix B.1.)

If you ask the question "Are you thinking of the number 7, that is, 0111?" we can quantify the information that is contained in your friend's answer. If she answers "Yes, it is," you received 4 bits of information: You know all four bits (0111) of your friend's answer. However, if your friend answers "No, it is not," there are 15 out of the 16 possibilities left, which means that you received very little information. How little, exactly?

In general, we define the *information content* of an event as

$$\log_2 \frac{1}{p},$$

where p is the probability of that event. Here, $\log_2$ is the logarithm to base 2 (appendix B.3 contains more details on the logarithm function). In the example above, this yields

$$\log_2 16 = 4 \quad \text{for the "yes" answer,}$$

$$\log_2 \frac{16}{15} \approx 0.093 \quad \text{for the "no" answer.}$$

The unit of information content is often also called a "bit." In our number guessing game, that name fits well.

As we may have expected, it is not a good idea to ask questions whose answers we already know: If the question is such that the answer has probability 1, the information content will be $\log_2(1) = 0$ bits. It is not helpful to ask questions such as "Is your number smaller than 1401?", for example, because the answer does not contain any helpful information—the information content equals zero bits.

In the opera game, there are not only 2 (yes/no) but also 3 possible answers (neutral/success/disaster). And if we consider two performances in a row, there are even $3 \times 3 = 9$ possible outcomes. But the key idea of information content remains the same. If there are m possible answers (with probabilities $p_1, p_2, \ldots, p_m$, say), then the information content of the m answers are, respectively,

$$\log_2 \frac{1}{p_1}, \log_2 \frac{1}{p_2}, \ldots, \log_2 \frac{1}{p_m}.$$

Information Content and Selecting Hypotheses

Information content is closely related to the ability to identify the true hypothesis among a set of possible hypotheses. This set is often written as Ω. Let us start again with the number guessing game described near the start of this section. There, $\Omega = \{0, 1, 2, \ldots, 15\}$.

We now describe any strategy as a map to help us navigate from the set of possible hypotheses to a set of possible answers. Let $\omega^* \in \Omega$ denote the secret number that your friend thinks about but that is unknown to us. Asking a yes/no question now corresponds to constructing a function ("O" for outcome),

$$O: \Omega \rightarrow \{0, 1\} \qquad \text{(one no/yes question)}$$

and asking your friend for the value of $O(\omega^*)$; that is, $O(\omega^*) = ?$. Here, we simply identify "yes" with 1 and "no" with 0. The question, "Are you thinking of the number 7?", for example, translates into a function O that maps 7 to 1 and all other values to 0. So, $O(\omega^*) = 1$ if ω^* happens to be 7, and $O(\omega^*) = 0$ otherwise. In this case, we have $P(O = 1) = 1/16$, because only one element in Ω maps to 1.

Asking two yes/no questions in a row corresponds to constructing a function $O: \Omega \rightarrow \{0, 1\}^2$. The set $\{0, 1\}^2$ of answers contains all possible combinations of yes and no answers: $(0, 0)$, $(0, 1)$, $(1, 0)$, and $(1, 1)$. Or, in words, (no, no), (no, yes),

(yes, no), and (yes, yes). Finally, asking k questions corresponds to constructing the function

$$O:\Omega \to \{0,1\}^k \qquad \text{(asking } k \text{ no/yes questions)}$$

and, again, asking your friend for the value of $O(\omega^*)$.

We thus think about the truth (that is, the secret number) as an element $\omega^* \in \Omega$ that is unknown to us. Asking questions or combinations of questions corresponds to constructing functions $O:\Omega \to \{0,1\}^k$. The answer, given by our friend, equals $O(\omega^*)$, the evaluation of the truth ω^*. This turns out to be a convenient description of the problem. The set Ω tells us the number of hypotheses, and the set $\{0,1\}^k$ tells us the number of possible answers. Now, for $k=3$, for example, the set $\{0,1\}^k$ contains 8 elements. But since Ω contains 16 elements, $\#\Omega = 16$, there are at least 2 elements in Ω that map to the same answer. That is, no matter what strategy we choose, we can never be absolutely certain that we will identify the correct number after $k=3$ questions. If, however, we are lucky, and the information content of a given answer $O=i$, say, is ≥ 4 bits, then we have identified the secret number ω^*:

$$\log_2\left(\frac{1}{P(O=i)}\right) \geq 4 \quad \Rightarrow \quad \frac{1}{P(O=i)} \geq 16 \quad \Rightarrow \quad P(O=i) \leq \frac{1}{16}.$$

Thus at most one element ω^* maps to the answer x (otherwise, the probability of obtaining that answer would be at least 2/16)—so this must be the true secret number ω^*! These two observations hold true in general, and we reiterate them here:

$$\#\Omega > \#\text{possible answers} \quad \stackrel{\text{Rule 1}}{\Longrightarrow} \quad \omega^* \text{ cannot always be identified,}$$

$$2^{\text{information content}} \geq \#\Omega \quad \stackrel{\text{Rule 2}}{\Longrightarrow} \quad \omega^* \text{ is identified.}$$

Rule 1 gives us a lower bound on the number of questions that we need to ask to be able to identify the truth even in the worst case. Rule 2 tells us that an answer with high information content allows us to identify the truth.

What about the opera game described in section 5.1? There, the set Ω contains 20 elements and is listed at the start of section 5.3. We will again describe strategies by using maps. But unlike for the number guessing game, we have 3 possible outcomes, so this time, any strategy can be described by

$$O:\Omega \rightarrow \{0,1,2\}^k \quad (k \text{ performances with neutral/success/disaster}).$$

We can now exploit the two rules that we learned about above. On one hand, $k=2$ performances do not suffice to always identify the true solution, because $\#\Omega = 20 > 9 = \#$possible aswers. On the other hand, receiving more than $\log_2(20) \approx 4.322$ bits of information content is in principle enough to identify the two stars and whether they are friends or enemies. This is because $2^{\log_2(20)} = 20 \geq \#\Omega$.

Because of the rules of the game (and unlike in the number-guessing game), we cannot implement any arbitrary function O as a strategy. For example, we cannot simply ask "Is this singer a star?" But the other direction works fine: Any strategy can be described as a function O.

In summary, receiving a lot of information content is good for identifying the correct hypothesis. In practice, we can control the information content by designing questions or experiments with certain outcome probabilities. We will now be looking at how to design the best questions (i.e., the ones that will prompt answers with a high information content). These answers will then help us identify the correct hypothesis.

Principle 1: Maximizing the Minimal Information Content

Let us assume that in the first opera performance, we put 3 of the 5 singers in the first ensemble and the remaining 2 in the second ensemble—all players are on stage. The probabilities of success and disaster will then both be 0.5 (each pair can yield either outcome, depending on whether they are friends or enemies). The neutral option will never come up. This means, we obtain the following bits of information:

Outcome	neutral	success	disaster
Information content	✗	$\log_2 2 = 1$	$\log_2 2 = 1$

Is this a good question to start with? It turns out that we can do much better in terms of information content. If instead, we leave 1 singer off-stage, there will be a neutral reaction in 8 out of 20 cases (namely, in the cases where the left-out singer is one of the stars). We would therefore obtain:

Outcome	neutral	success	disaster
Information content	$\log_2(20/8) \approx 1.32$	$\log_2(20/6) \approx 1.74$	≈ 1.74

bits for the neutral, success, and disaster outcomes, respectively. Every outcome of the second performance will therefore contain more information than any outcome of the first performance.

The theory of information provides us with a powerful tool for designing questions. If we want to minimize the maximal number of questions required to find out the truth, we need to ensure that even in the worst case, the answer's information content is as large as possible. Let us assume that there are m possible outcomes, and the probability $P(O=i)$ of receiving outcome i equals p_i. The outcome with the smallest information content is the one corresponding to

$$\min_{i \in \{1,\ldots,m\}} \log_2 \frac{1}{p_i},$$

that is, the one corresponding to the largest probability p_i. This yields the following rule:

We maximize the minimal amount of information by making the most probable answer as unlikely as possible (that is, by making the largest p_i *as small as possible).*

Principle 2: Maximizing the Average Information Content

Another idea is to find a question whose answer contains a lot of information *on average*. Consider a question that is answered "yes" with probability p and therefore "no" with probability $1-p$. The expected information is then defined as

$$H(p) = p\log_2(p) + (1-p)\log_2(1-p),$$

and is usually referred to as the *entropy*. We adapt the widely used convention of defining $0 \cdot \log_2 0 := 0$.[2]

In the example of the number-guessing game, the entropy of our question "Are you thinking of the number 7, that is, 0111?" equals

$$\frac{1}{16}\log_2 16 + \frac{15}{16}\log_2 \frac{16}{15} \approx 0.337,$$

that is, we receive 0.337 bits of information on average. Most people would not start with the above question, but would rather ask: "Is your number greater than or equal to 8?" for example. The entropy clarifies in which sense this question is more informative. On average,[3] the latter question yields

$$\frac{1}{2}\log_2 2 + \frac{1}{2}\log_2 2 = 1$$

bit of information. This makes sense intuitively: The answer to "Is your number greater than or equal to 8?" tells us the first bit in the binary expansion of the secret number.

In the general case of m possible answers, we define the *entropy* as the expected information content:

$$H := H(O) := H(p_1, \ldots, p_m) := p_1 \log_2 \frac{1}{p_1} + \cdots + p_m \log_2 \frac{1}{p_m}.$$

2. This convention makes sense, as $p \cdot \log_2 p$ is converging to 0 if $p > 0$ is converging to 0.

3. In fact, the minimal information content is also 1 bit in this case.

Appendix B.6, "What Is ... an Expectation?" contains a few more details on why this is the *expected* information content. In many games, we are not interested in worst-case behavior, but we would like to receive a lot of information on average. To do so, we should design the question so that its entropy is as large as possible. But how do we achieve that? In short, the entropy is maximized if all m answers are equally likely, that is, $p_1 = p_2 = \cdots = p_m = 1/m$. A proof of that statement can be found in appendix C.3. This observation yields the following rule:

> *The average amount of information is maximized if all* p_is *take the same value.*

What does that mean, intuitively? If all answers to a certain question are equally likely, the average information content of the answer is as large as possible. The other extreme is if one of the m answers has probability 1 and all others have probability 0. Then, the average information content is 0.

If it is possible to find a question such that all answers have the same probability, both the average information content and the minimal information content are maximized and equal. This is what we should aim for! Often, however, finding questions with equal answer probabilities is impossible, and the two concepts differ from each other.

Finding Optimal Strategies

With these principles in mind, we can now try to construct useful strategies. In the opera game, there are three outcomes, so we should design a sequence of performances, represented by the map

$$O : \Omega \rightarrow \{0, 1, 2\}^k$$

such that all the outcomes have (roughly) equal probability. This, however is a difficult task. In practice, we can often design the performances one-by-one. Starting with $O_1 : \Omega \to \{0, 1, 2\}$, we try to design the performance in such a way that each outcome has probability 1/3. In our solution on page 78, the probabilities equal 6/20, 6/20, and 8/20 and are as close as possible to 1/3, 1/3, and 1/3. But what happens to the information if we add a second performance? The amount of information content that the second answer adds to the first answer depends on how related the two questions are. Assume, for example, that in the number-guessing game, we ask two yes/no questions: "Is your number even?" and "Does the last digit in the binary expansion equal one?". Both of these questions individually receive on average 1 bit of information ($1 = 1/2 \log_2 2 + 1/2 \log_2 2$). But both of the answers together are still worth 1 bit of information. The reason is that the two answers depend on each other: If the first question is answered with "yes," then the second answer must be "no" and vice versa. If, however, the questions are constructed in such a way that the answers are independent (e.g., "Does the first digit equal 1?" and "Does the second digit equal 1?"), the average amount of information of the combined questions is maximized and equals the sum of both entropies: 2 bits. This is not too hard to prove—see appendix B.5 for details on independence.

No matter what the outcome of the first performance, we thus need to construct a second performance whose outcome is as independent as possible from that of the first performance. In practice, this yields the following rule:

> *At each point in time, ask a question the answers to which, given current knowledge, have roughly equal probability.*

In fact, the solution shown on page 78 was constructed by such a step-wise procedure. At each point in time, we constructed the performances in such a way that any outcome had roughly equal probability.

This procedure is simple and can be applied to many games, where one needs to identify a hypothesis (the section 5.5 discusses another game of this type). However, there is a drawback to the stepwise approach: It does not always yield an optimal strategy (we provide an example in appendix C.3). In our case, however, we are lucky. We know from before (page 83) that there is no strategy based on only two performances or fewer that can always identify the true hypothesis. Since our strategy involves three performances, it must be optimal.

The idea of maximizing information content not only helps with mathematical problems. We usually consider the conversations with high information content as particularly interesting. Questions whose answers we can guess with high probability are (on average) boring. But in a few cases, they yield high information content. For example, asking a stranger about a friend you may have in common usually results in a negative answer; but if it does not, the surprise is large—and so is the information content.

5.5 VARIATION: BALL WEIGHING

The opera singers problem is a variation on a famous ball-weighing problem. That problem is formulated as follows. There are 12 balls, 11 of which are of equal weight. The twelfth ball, however, is different; it weighs a different amount, but that difference is unknown: It is either lighter or heavier than the other balls. Since the odd ball out is the same size and color as the other balls, it is visually indistinguishable from the others. The task is now to use a balance scale to identify the "oddball" with as few weighings as possible (see the figure on the next page). Here, the balance only indicates whether both sides are equally heavy and, if they are not, which side is heavier.

The ideas we discussed in this chapter can also be used to solve the weighing problem. Which weighing shall we start with? Do we expect to gain more information from weighing 6 against 6 balls, 4 against 4, or 1 against 1? Section 5.7 tells you where to find the solution.

5.6 RANDOM STRATEGIES

For most problems, designing strategies that optimize information content is difficult. We have seen that the stepwise approach can be helpful in practice but does not necessarily provide the best strategy (see appendix C.3). However, there is a simple and surprisingly useful alternative: We can simply design the experiments randomly. In the ball-weighing problem, for example, we can put 4 randomly chosen balls on the left-hand side of the balance and 4 other randomly chosen balls on the right-hand side. After 3 such random weighings, we can then check how many hypotheses could be used to explain the results. We repeated this experiment 10,000 times (fortunately, we were able to use a computer) and found that in about 47% of the cases, only a single hypothesis was left. If more than one hypothesis is left (e.g., ball 3 could be lighter or ball 9 heavier than the other balls), we can also randomly choose one such hypothesis (e.g., we guess "Ball 9 is heavier than the other balls"). This allows us to correctly identify the special ball in about 72% of cases. For 4 weighings, these numbers increase even further to 81% and 90%, respectively.

This random strategy is extremely simple and fast. Each weighing is randomly constructed without the need to adapt to the result of the previous weighing. When considering a large number of balls, for example, finding the optimal strategy is very difficult—even the stepwise procedure is a bit cumbersome to compute. The random strategy, however, remains simple. Motivated by our deliberations on information content and entropy, we always choose $\approx p/3$ randomly selected balls on each side of the balance, where p is the number of balls. The following graph shows how many weighings are needed to obtain a success probability of greater than or equal to 90% (again allowing for random guesses if there are ties). For $p = 2{,}048$ balls, we only need 9 weighings!

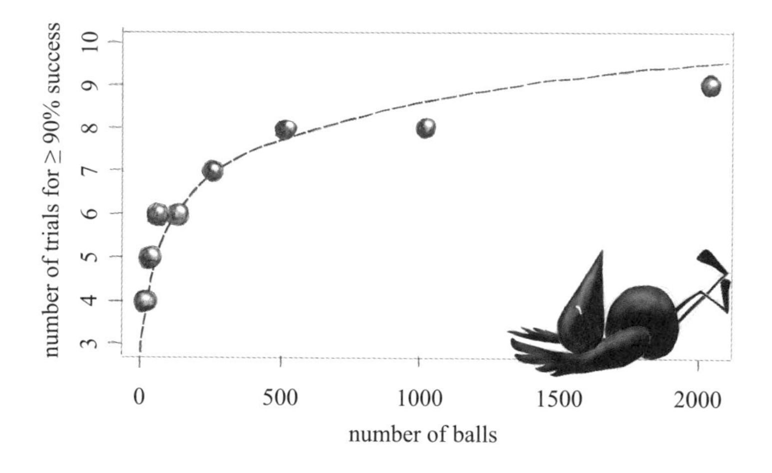

It seems surprising that the number of necessary weighings grows so slowly with the number of balls. But there is mathematical theory that explains this behavior. It tells us that for a growing number p of balls, we only require (roughly) $C \cdot \log(p)$ weighings. The figure on the previous page shows a graph of that function for a specific choice of C: It fits the black dots quite nicely. The underlying theory is sometimes called "compressed sensing" or "compressive sampling." To apply it to our problem, we represent the truth by a long vector β of length p that contains only 0s, apart from at the position of the secret ball, where the vector has either a $+1$ or a -1, indicating whether the ball is heavier or lighter than the others. A weighing can now be described by a vector x of length p, for which the entries -1, 1, and 0 indicate, respectively, that the corresponding ball is on the left-hand side of the balance, on its right-hand side, or is not included. The result of a weighing then corresponds to the measurement $\sum_{i=1}^{p} x_i \beta_i$. The values -1, 1, and 0 indicate that the left-hand side is heavier, the right-hand side is heavier, or that they are equally heavy, respectively. The theory of compressed sensing now tells us that the number of measurements we need to reconstruct the vector β, grows only like the logarithm of p. One of the reasons the random strategy works so well is that out of all these balls, only one ball differs from the others. Random strategies still work if there are only a few balls that are different from the majority, but if there were lots of balls like this, we would have no chance of recovering them with only a few random weighings. The theory of compressed sensing can not only be used for our weighing problem. It can also be used in many other areas, such as seismology, photography, and medical imaging.

5.7 SHORT HISTORY

To the best of our knowledge, the earliest formulation of the weighing problem appeared in a 1945 article in the *American Mathematical Monthly* [Schell, 1945], with a more general game published the following year [Eves, 1946]. You can read more about the ball-weighing problem and its solution (there is a strategy that uses at most 3 weighings) in the book *Information Theory, Inference, & Learning Algorithms* [MacKay, 2002] on page 68, for example. Our figure on page 78 is inspired by David MacKay's figure on page 69 of his book. The concept of entropy [e.g., Cover and Thomas, 2006] and its connection to information content has been studied by Claude Shannon, who was one of the pioneers in that area [Shannon, 1948]. His original motivation was to study communication, and today the theory is widely applied not only in communication and signal processing but also in data compression and data science, as well as in other areas. The theory of compressed sensing [Candès et al., 2006; Donoho, 2006] was developed by Emmanuel Candès, David Donoho, Justin Romberg, and Terence Tao.

5.8 PRACTICAL ADVICE

In a first instance, the game can be played competitively. The host divides the players into three groups in each round of the game. She can do the division in any way she likes and declares the performance again either a success, disaster, or neutral, according to the same rules. The players have to guess who the stars are and what their relationship is. The first player to get it right wins the game (one can allow for one false answer per person). Once the players have played this competitive round, they can move to the cooperative version of the game, where they have to agree on the division of the group and guess the identity of the stars and their relationship together. The players should aim to be faster (i.e., to use fewer rounds) than in the competitive round of the game.

6 ANIMAL MATCHING AND PROJECTIVE GEOMETRY

6.1 THE GAME

Number of players:	7 (see section 6.6 on advice for using different numbers of players)
You will need:	7 different kinds of animal toys, 3 of each animal

The 21 different toy animals are arranged on a table; 3 lions, 3 mice, 3 snakes, 3 koalas, 3 cats, 3 elephants, and 3 giraffes. The players enter the room one after another, and each player takes 3 animals. Players are not allowed to see what the other players have chosen. The players can hide their choices behind their backs or in their pockets, so neither the other players nor the audience can see which animals they have taken.

The audience is now asked to select a pair of players. The players' challenge is to produce a common animal from the 3 they have behind their back. Say the first pair of players are as shown on the previous page. Once the pair of players is chosen, they present all the animals they have taken and see whether they find a match between them. They have both taken a giraffe and will each show the giraffe to the audience to answer the challenge. If they do not have an animal in common, all players lose. If they do produce a match, the audience can select a new pair of players, challenging them to produce a matching pair of animals. The audience can ask 3 times, and the players only win if all 3 pairs of players can show the same animal.

To make the game more difficult, the audience can, for each player, choose the first of the 3 animals that the player has, but they must make sure that they choose a different animal for each person.

What is the best strategy for the players? What is the best strategy for the audience? How likely are the players to succeed? What about other numbers of players and animals?

We can check the success rate of some simple strategies. The simplest strategy is perhaps that all players take animals at random from the ones remaining on the table. In the case of 7 players, the chance of success of the team will be just over 27%.

There is an obvious way to improve the purely random strategy. For a given player, it is clearly not optimal to have 2 animals of the same kind in her hand. A simple modification is hence that each player makes sure that she takes three distinct animals from the ones remaining (at least as long as possible; the players toward the end of the selection process might no longer have the option of taking 3 distinct animals). This simple modification already improves the success rate considerably. Let us first look at 2 randomly chosen players out of the 7. If both players take 3 animals at random from those available, the chance of a match between them will be around 64%. If they take 3 distinct animals at random each, then they have a chance of just over 88% of having at least 1 animal in common. If all the players make sure that they each take 3 different animals from those remaining, for as long as it is possible to take 3 different animals, their chances of success will already be almost 63% instead of the 27% with the previous strategy. Can the team beat the 63% success rate?

Another way to think about the problem is to first just look at the koalas. There are 3 koalas in the game. Let us assume that the 3 koalas are distributed among 3 different players. Then each of the 3 possible pairs of these 3 players will be "linked," as they possess an animal in common (the koala). So the koala is introducing 3 links in the game, where a *link* is a pair of players that can successfully show an animal in common. Each of the other animals is also introducing 3 links into the game (and fewer than 3 if players start having more than 1 of each kind of animal). The 7 animals in total thus introduce 21 links. How many links would we need to make sure that each pair of players has an animal in common? We would need as many links as there are pairs of players. For 7 players, we have $7 \cdot 6/2 = 21$ unique pairs. So the number of available links just about matches the number of links we need to connect all players! Therefore, it seems just about possible that the game could be played with a 100% success rate. But we cannot waste any link if we want to achieve this. If one pair of players has 2 links in common (they share more than one pair of animals), then, as a consequence, there would have to be another pair of players with no animal in common. We believe that now is a good time for the reader to pause and think about a possible solution.

6.2 SOLUTION

There is one solution that will never fail, because every pair of players the audience selects is guaranteed to have an animal in common (that is, the success rate equals 100%). If the players choose their 3 animals following a pattern equivalent to the one set out below:

Animal assigned by audience	Other animals taken
lion	cat, mouse
mouse	koala, snake
snake	giraffe, lion
koala	giraffe, cat
cat	elephant, snake
elephant	lion, koala
giraffe	elephant, mouse

If you take any pair of players with this choice of animals, they will always have 1 animal in common.

Is it also possible to construct such a solution if we have 8 players and 8 different kinds of animal? In general, it turns out that solutions are known if p is a prime number (or a power of a prime number) and there are p^2+p+1 players and p^2+p+1 unique animals, with each player taking $p+1$ different animals. In our example, $p=2$ and therefore, we have 7 players and 7 unique animals, with each player taking 3 animals. For $p=3$, we play with 13 players and 13 unique animals, with each player taking 4 unique animals.

6.3 FANO PLANES

Let us look at the case of $p=2$ and 7 animals. Each of the 7 players takes 3 animals.

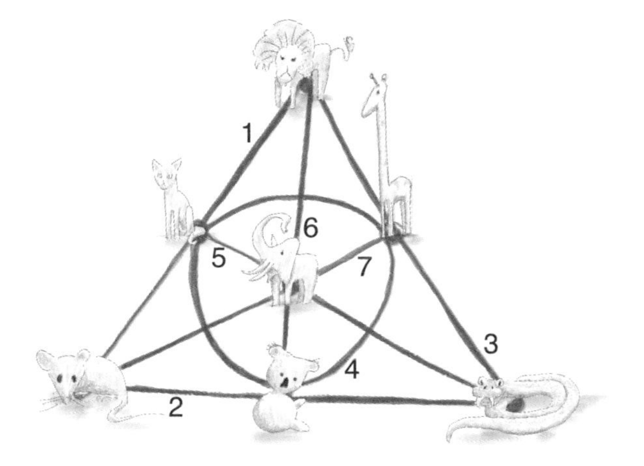

The solution can be characterized by the so-called Fano plane. Each point corresponds to 1 of 7 animals. In the above figure, each of the lines marked 1–7 corresponds to 1 player. Here, we call line 4 a line, even though it is bent. Each line (player) passes through exactly 3 points (animals). Each point (animal) is contained in 3 lines (players).

The way the animals are shared among the players can be recorded in a so-called incidence matrix A that contains only 0s and 1s. Each row corresponds to a player, and each column corresponds to an animal. In our example, the matrix A has dimension 7×7 for $p = 2$, as we have 7 players and 7 animals. A value of 1 in position (k, j) indicates that player k has taken animal j.

	Lion	Mouse	Snake	Koala	Cat	Elephant	Giraffe
Player 1	1	1	0	0	1	0	0
Player 2	0	1	1	1	0	0	0
Player 3	1	0	1	0	0	0	1
Player 4	0	0	0	1	1	0	1
Player 5	0	0	1	0	1	1	0
Player 6	1	0	0	1	0	1	0
Player 7	0	1	0	0	0	1	1

$= A.$

The first row of A indicates that player 1 takes a lion, a mouse, and a cat. This corresponds to line 1 in the Fano plane connecting the lower left with the top corner of the triangle. The second player (second row in A) takes a mouse, a snake, and a koala, corresponding to line 2 in the Fano plane.

We show in appendix C.4 how the desired properties of the solution (every pair of players has an animal in common) relate to properties of the incidence matrix. The relation will also help us understand why every player has to hold exactly 3 animals; there is no valid solution where one player, for example, holds only 2 animals and another players holds 4.

So, the question is how to construct general incidence matrices that can also be applied to other numbers of players and animals.

6.4 SOME MATHEMATICS: PROJECTIVE GEOMETRY

In section 6.3, we have identified lines in a plane with players and points on the plane as animals. In mathematics, the relation between points and lines is studied in the field of geometry. Formally, a line is just a set of points; we say that the point is *incident* with the line and the line is *incident* with the point if the point lies on the line.

For a standard 2-dimensional Euclidean plane (think about drawing points and straight lines on this page), the following two statements are valid:

I. Given any 2 distinct points m and m', there is exactly 1 line ℓ incident with both of them.

II. Given any line ℓ and any point m not incident with ℓ, there is exactly 1 line incident with m that does not meet ℓ.

Property I states that any two points can be connected by exactly one line. The line in question in property II is, of course, a line that is parallel to ℓ, as parallel lines do not intersect in standard Euclidean geometry. If we associate lines with players and points with animals, then property II can be problematic, as it means that 2 players who correspond to parallel lines have no animal in common.

Projective geometry is different from the Euclidean geometry that we are used to. In a *projective plane*, every pair of lines will intersect at exactly 1 point (which can, perhaps unintuitively, be achieved by adding a point at infinity for each pair of parallel lines), so that the following statements hold:

i. Given any 2 distinct points, there is exactly 1 line incident with both of them.

ii. Given any 2 distinct lines, there is exactly 1 point incident with both of them.

These are the properties we are looking for. Assume that we have just a finite number of points. Then we will have a finite number of lines, and the properties i and ii are exactly what we desire: Each pair of animals will appear for exactly 1 player, and as desired in the game, each pair of players will have exactly 1 animal in common.

A Simple Construction of the Solution

An illustration of a solution for $p=3$ with $p^2+p+1=13$ distinct animals and 13 players is shown on the next page. The animal types are identified with the numbers 1, 2, ..., 13. The numbers 1 to $p^2=9$ are arranged in a square, while the remaining numbers 10, 11, 12, and 13 indicate different slopes of lines passing through this grid of points.

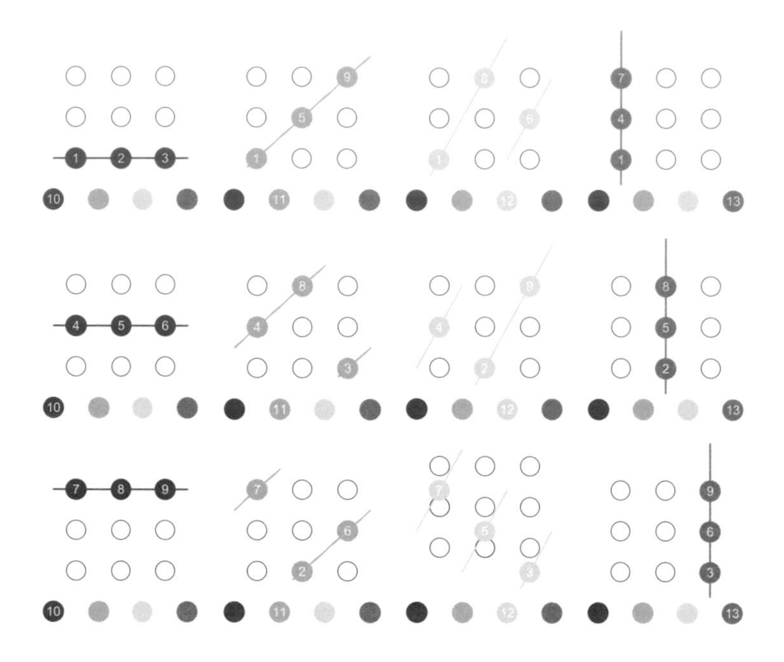

Number 10, for example, corresponds to the slope of all horizontal red lines and can be interpreted as the "point at infinity." In this geometric construction, the nodes in both dimensions of the $p \times p$ grid are defined modulo p (see chapter 4). The modulo operation can be visualized by multiple copies of the original $p \times p$ grid, arranged to the left, right, top, and bottom of the original $p \times p$ grid. The lines shown above in the original grid are straight lines in this extended version.

A player corresponds again to a line; she takes the animals of the 3 numbers that her line passes through and also the animal that corresponds to the slope indicator of her line. The top-left figure panel corresponds to the first player; she holds animals $1, 2, 3$ and also animal 10 as an indicator for the horizontal line that is used to connect $1, 2,$ and 3. The players in the left column of panels in the figure all have a horizontal slope and hence all have the slope indicator 10 in common. Lines with a different slope will intersect in exactly 1 animal on the square grid. We have p^2 animals on the grid and $p+1$ slopes, so there are p^2+p+1 distinct animals. The figure on the previous page shows $p(p+1)$ combinations of points on the grid and slopes. The configuration for the last player (not shown) consists of all the slope indicators 10, 11, 12, 13. This last configuration will also have exactly 1 animal in common with all of the configurations shown above. The slope indicators force parallel lines to intersect in the sense of properties i and ii. The construction works only if p is a prime (or power of a prime), as lines with different slopes would otherwise not necessarily intersect.[1]

A More Formal Construction

The role of the prime numbers will become more apparent in a more abstract construction, which will lend itself more easily to actual implementation. Say we start with a prime number p and the set of integers $\{0, 1, \ldots, p-1\}$. For $p=2$, the set is $\{0, 1\}$, for example. Then we can define addition and multiplication by taking the results modulo p (this defines a so-called *finite field*; see chapter 4 for more discussion of the modulo operation and cyclic groups). If we now look at a 3 dimensional vector space over this field, we get for $p=2$ the following 8 points:

$$(0,0,0), (0,0,1), (0,1,0), (1,0,0), (0,1,1), (1,1,0), (1,0,1), (1,1,1)$$

as the 8 corners of a unit cube. Taking the results of addition and multiplication modulo 2 means that, for example,

$$(0,0,1)+(0,1,1)=(0,1,0).$$

Now one can look at the set of all lines in the 3-dimensional vector space that pass through the origin and one of the other

1. You can try this, for example, on a 6×6 grid.

points. For $p=2$, there will be 7 distinct lines, each passing through the origin and one of these points:

$$(0,0,1), (0,1,0), (1,0,0), (0,1,1), (1,1,0), (1,0,1), (1,1,1).$$

Now we can identify each line passing through the origin with an animal:

(1) lion:	$(0,0,0) - (0,0,1)$
(2) mouse:	$(0,0,0) - (0,1,0)$
(3) snake:	$(0,0,0) - (1,0,0)$
(4) koala:	$(0,0,0) - (1,1,0)$
(5) cat:	$(0,0,0) - (0,1,1)$
(6) elephant:	$(0,0,0) - (1,1,1)$
(7) giraffe:	$(0,0,0) - (1,0,1)$

The line passing through $(0,0,0)$ and $(0,0,1)$ thus corresponds to a lion and the result is illustrated below.

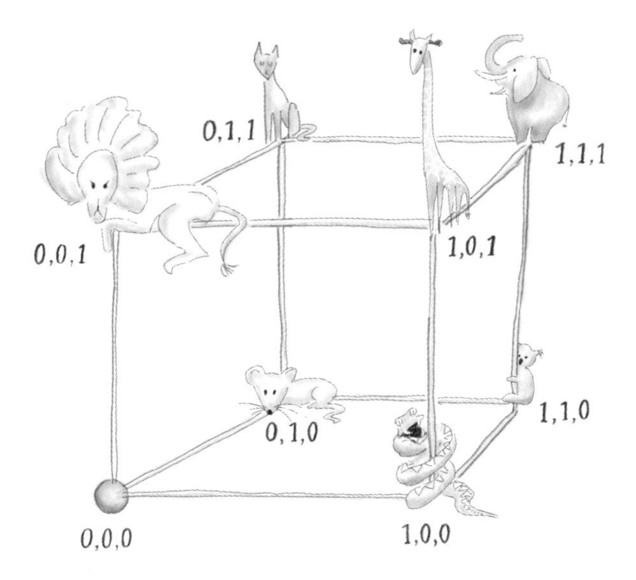

Therefore, the lines in the 3-dimensional vector space are the animals, (i.e., the Fano points), but what are the players (i.e., the Fano lines)? These will be the planes in the 3-dimensional space!

These planes are obtained by taking two lines and then considering all the sums of all the points on such lines. If we know two of the animals already, we will be able to determine what the third animal will be. For example, the plane that contains both lion and cat is passing through the points

$$(0,0,0), (0,0,1), \text{ and } (0,1,1).$$

Since $(0,0,1)+(0,1,1)=(0,1,0)$, the plane will also contain the line $(0,0,0)-(0,1,0)$ and hence a mouse. In this sense we can write

$$\text{lion} + \text{cat} = \text{mouse},$$

and the player, holding a lion and a cat, will also have to hold a mouse.

There are 7 distinct planes in the 3-dimensional vector space and we identify them with players in the game (or Fano lines). We have

$$\text{player 1: plane through } \{(0,0,0), (0,0,1), (0,1,0), (0,1,1)\}.$$

As we have seen, the plane of player 1 contains the animals lion, mouse, and cat. Note that any combination of the points on this plane will again be one of these four points, as we have to take the result modulo 2, so that, for example,

$$\begin{aligned} 2\cdot(0,0,1)+(0,1,0)+(0,1,1) &= (0,0,0)+(0,1,0)+(0,1,1) \\ &= (0,0,1), \end{aligned}$$

which can be expressed as

$$2\cdot\text{lion} + \text{mouse} + \text{cat} = \text{lion}.$$

For player 2, we have

$$\text{player 2: plane through } \{(0,0,0),\ (0,1,0),\ (1,0,0),\ (1,1,0)\},$$

which contains the animals mouse, snake, and koala. Next is

$$\text{player 3: plane through } \{(0,0,0),\ (0,0,1),\ (1,0,0),\ (1,0,1)\},$$

corresponding to the lion, snake, and giraffe, and so on.[2]

For $p=3$, we start with the set $\{0, 1, 2\}$, instead of $\{0, 1\}$ for $p=2$, and now take results of additions and multiplications modulo 3. The 3-dimensional vector space will then have 3^3 unique points:

$$(0,0,0), (0,0,1), (0,0,2), (0,1,0), (0,1,1), (0,1,2), (0,2,0), \ldots.$$

We can now enumerate the lines passing through the origin. This can be done, for example, by considering all points whose first nonzero coordinate is a 1:

$$(0,0,1), (0,1,0), (0,1,1), (0,1,2), (1,0,0), (1,0,1), (1,0,2),$$
$$(1,1,0), (1,1,1), (1,1,2), (1,2,0), (1,2,1), (1,2,2).$$

The corresponding lines are constructed by taking multiples of such points; for example, $(0,0,0)$, $(0,1,2)$, and $(0,2,1)$ form one line. These 13 points generate all unique 13 lines through the origin.[3]

Planes are generated by linear combinations of two lines. A plane, for example, that contains the origin, $(0,0,1)$, and

2. Note that when considering the cube as a subset of Euclidean space, not all 2-dimensional planes correspond to a player. For example, (koala, giraffe, cat) corresponds to a player as koala + giraffe = cat according to the calculus we have described, but (lion, snake, mouse) does not correspond to a player as lion + snake = giraffe.

3. This can be seen as follows: The lines are all distinct, since any multiple of any of the listed points cannot result in another point in the list. Furthermore, for any nonzero point v whose first nonzero coordinate is not a 1, there exists a $\lambda \in \{1,2\}$, such that λv is contained in the list; because $v=\lambda^{-1}\lambda v$, the point v lies on the line generated by λv.

$(0,1,0)$, will also contain $(0,1,1)$ and $(0,1,2)$. It will also contain the line through the origin and $(0,0,2)$, for example, but this line is identical to the line through $(0,0,1)$. The construction is then analogous to the case of $p=2$ by identifying planes with players and lines with animals, although these roles can also be reversed.

In general, as long as p is a prime number, this 3-dimensional embedding yields p^2+p+1 distinct lines (using the same enumeration of lines as above) corresponding to animals and p^2+p+1 unique planes corresponding to players. Each plane will contain $p+1$ lines, and each line is contained in $p+1$ distinct planes. Formal proofs that this construction does indeed yield a projective plane with the required number of lines and points can be found in [Kåhrström, 2002], for example.

The construction is also possible for prime powers (for example, if p equals $4=2^2$ and $8=2^3$ and $9=3^3$) but this is a bit more demanding, as we cannot use the integers modulo p as a field any longer, and we have to construct suitable polynomials instead. Hence the values $p=2,3,5,7,11,\ldots$ are all covered by the construction above, and by an extended argument, we can also cover the prime powers $p=4,8,9,16,\ldots$. The four smallest numbers in $\mathbb{Z}$ that are neither prime nor prime powers are 6, 10, 12, and 14. The Bruck-Ryser theorem states that there is no solution if p is such that it is congruent to 1 or 2 modulo 4 and is not a sum of two squares. Thus, there is no solution for $p=6$ and $p=14$, for example. The theorem does not say anything about the cases $p=10$ and $p=12$, however. While it has been shown (using a computationally very demanding proof) that there is no solution for $p=10$, what happens for $p=12$ or other numbers not covered by the above characterizations is still open.

6.5 SHORT HISTORY

The game "Dobble" (also called "Spot It!") is an example of already-constructed incidence matrices for $p=7$. There are 55 cards, showing 8 out of 57 unique objects. Each player is given a pile of cards and a starting card is placed in the middle.[4] Looking at the top card on their respective piles, each player can put it on the central pile as soon as he spots the object in common between his topmost card and the topmost card on the central pile. The first player to get rid of all of his cards wins the game. So, the aim of the game is to spot the common unique object. A "Mini Dobble" game with $p=2$ (that is, 7 cards with 3 objects each) was created by Maxime Bourrigan. A nice overview article appeared in *Math Horizons* in 2015 [Polster, 2015]. The simple construction of the first solution in section 6.4 is based on an answer by Sven Zwei on stackoverflow.com. The Bruck-Ryser theorem can be found in Bruck and Ryser [1949]. Lam [1991] showed that there is no projective plane for $p=10$. Projective planes are also used in experimental design; see, for example, Hughes and Piper [1985].

6.6 PRACTICAL ADVICE

There is a variation of the game that does not require the players to have such a good memory of the Fano plane as in the version described earlier in this chapter. The game is for one less player than in the original version, that is, 6 players with 7 unique animals. The audience is allowed to remove 2 animals at the start. Then, each player can take 3 of the remaining animals. For

4. The card in the middle is the reason that the total number of cards is 55 and not $p^2+p+1=57$. With 55 cards in total, the remaining 54 cards can be distributed equally among either 2 or 3 players.

this variation, a viable strategy is that each player memorizes one line in the table below, and all players remember the last line of the table as a backup option.

player 1	lion	mouse	cat
player 2	mouse	snake	koala
player 3	lion	snake	giraffe
player 4	cat	giraffe	koala
player 5	cat	snake	elephant
player 6	lion	koala	elephant
backup option	mouse	giraffe	elephant

If a player sees that two of the animals in his primary line have been taken away by the audience, then he can switch to the backup option. Then all the pairs of players will once again have 1 animal in common, as each pair of animals uniquely identifies a line.

For 12 players and 13 unique animals, we can use the table below.

player 1	lion	mouse	cat	snake
player 2	lion	giraffe	koala	spider
player 3	lion	elephant	eagle	earthworm
player 4	lion	frog	dolphin	jellyfish
player 5	mouse	giraffe	elephant	dolphin
player 6	mouse	koala	eagle	frog
player 7	mouse	spider	jellyfish	earthworm
player 8	cat	giraffe	eagle	jellyfish
player 9	cat	koala	dolphin	earthworm
player 10	cat	elephant	frog	spider
player 11	snake	giraffe	frog	earthworm
player 12	snake	koala	elephant	jellyfish
backup option	snake	eagle	spider	dolphin

The game can also be turned around, and the audience could be allowed to ask for any pair of animals, to see whether there is a player who has taken this particular combination. This would also always be true under these strategies, and the players would cover all possible animal-pairs among themselves.

The types of the animals can, of course, be varied or replaced by fruits, stickers, or children's names.

7 THE EARTH AND AN EIGENVALUE

7.1 THE GAME

Number of players:	2 or more
You will need:	2 (inflatable) world globes

Two contestants are each given an inflatable globe. They are asked to place the globe on the floor. The globe may be in any rotation that they choose. Once they have placed the globes, the goal is to find a spot on the globe, such as a city, that has the same position on both globes. Being in the same position means that if 2 tiny people would stand on that place on each of the inflatable globes and would point a laser pointer into the sky, perpendicular to the surface, the 2 laser beams would be parallel. The place can be anywhere on the globe, and it can be a city, a point in a desert, or in the ocean. As an example, if both contestants chose to place the South Pole on the floor, then the North Pole would be such a point: on both globes, it points vertically into the sky (the South Pole would be another correct answer, of course). But the contestants can choose any rotation, so in general, the correct point does not need to be the North Pole, and it does not need to point vertically upward. The figure below shows another example.

The task is now to find the place on the globe that has exactly the same position on both globes. In this case, it is a location close to Seattle, marked by a flag on both globes in the figure (which is just for illustration; the flag would not be present in the actual game).

The person who is first to announce such a spot on the globe (or, alternatively, declaring that no such place exists) wins the game. But only, of course, if the answer is correct. If it is incorrect, the other person wins. Even though the game is about being fast, we do not discuss strategies on how to find such a point. Instead, we turn to a more fundamental question.

Does such a point always exist? And if so, how can we prove that it exists?

We can view the left globe as the reference globe and the right one as a rotated version. The question becomes whether there is a point on the right globe that has exactly the same position as on the reference (left) globe.

Mathematically, the globes can be described as sets of points in 3-dimensional space. And a rotation R is a map: Every point on the left globe is mapped to another point on the globe (to obtain the right globe).

Let us start by describing the left globe mathematically. Its placement can be arbitrary, but for simplicity we assume that it is placed such that the North Pole is pointing upward and the prime meridian "longitude 0°" (this is the line on which Greenwich, UK, lies and that also passes through Ghana and Togo) is in the front (see the picture on the next page). We describe the center of the left globe as

$$\text{center of globe:} \quad v = \begin{pmatrix} 0 \\ 0 \\ 0 \end{pmatrix}.$$

For simplicity, let us further say that each of the 2 globes is perfectly round, and that it has radius 1 (e.g., 1 meter). Each point on the globe can then be represented as a 3-dimensional vector:

$$\text{point on globe surface:} \quad v = \begin{pmatrix} v_1 \\ v_2 \\ v_3 \end{pmatrix} \quad \text{with } v_1^2 + v_2^2 + v_3^2 = 1,$$

where the latter condition ensures that v has distance 1 from the center. Considering the above orientation of the left globe, the North Pole can be described by the coordinates

$$\text{North Pole (left globe):} \quad v = \begin{pmatrix} 0 \\ 0 \\ 1 \end{pmatrix}$$

and all points on the prime meridian, such as Greenwich have a 0 y-coordinate (see the figure below).

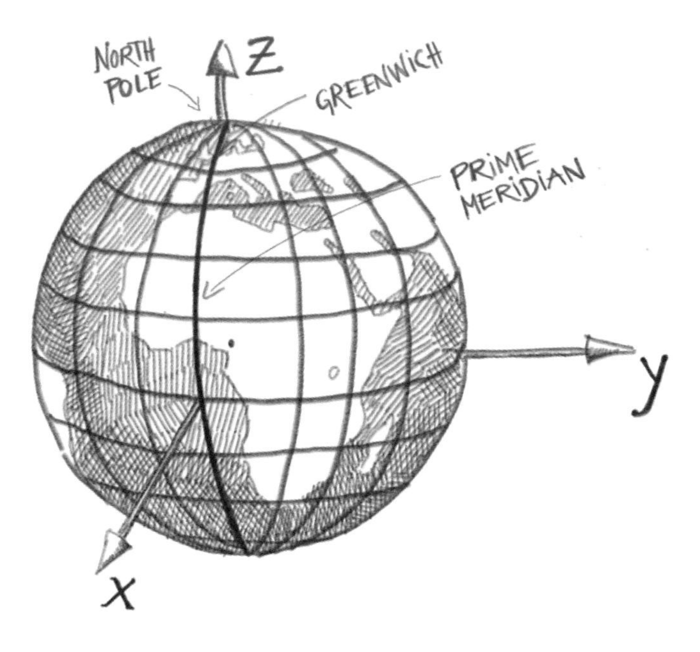

Having a model for the globe, we can now concentrate on the map R. If the rotation R (which can be a sequence of rotations) is such that the North Pole is mapped to the position where the South Pole was, then

$$R\begin{pmatrix} 0 \\ 0 \\ 1 \end{pmatrix} = \begin{pmatrix} 0 \\ 0 \\ -1 \end{pmatrix}.$$

Alternatively, if the rotation is such that Zurich is mapped to where Copenhagen was before, we would have

$$R(\text{Zurich}) = \text{Copenhagen},$$

that is, roughly,[1]

$$R\begin{pmatrix} 0.6722088 \\ 0.1010620 \\ 0.7334315 \end{pmatrix} = \begin{pmatrix} 0.5528803 \\ 0.1232626 \\ 0.8240932 \end{pmatrix}.$$

In this game, we are looking for a location v that, after the rotation, does not change its position:

$$R(v) = v.$$

We encourage the reader to pause here and think a bit more about the problem and whether such points exist.

7.2 SOLUTION

The surprising aspect of the game is that there are always at least 2 points on the globe that share the same position on the 2 globes. More precisely, either there are exactly 2 positions on opposite sides of the globe that share the same position, or the 2 globes are in exactly the same position, so all points are identical. It is thus always wrong to declare that there is no place with identical positions on the 2 globes, and the players have to keep looking until they spot the correct places (which can be surprisingly difficult if playing for the first time).

Mathematically, no matter what R is, there are at least 2 points v that satisfy $R(v) = v$. And they lie on opposite sides of the globe: if v remains fixed, then $-v$ does so, too. More precisely, there are either 2 such points, or we have $R(v) = v$ *for all* points v. In the remainder of this chapter, we will prove the above statement. To do so, we need to study properties of rotations.

1. These coordinates are sometimes called "ECEF" coordinates: earth-centered, earth-fixed; the system considers a y-axis pointing to the right.

Consider first a *basic rotation* of the globe in 3-dimensional space: It consists of first choosing an axis of rotation and, second, rotating the object by a certain angle. Angles can be measured in different ways. For example, as a number between 0° and 360° or between 0 and 2π. We will follow the preference of many mathematicians and use the second notation. The angle is then equal to the distance one would have to walk on the circumference of a circle with radius 1 to complete the rotation. A *rotation* we define as a concatenation of such basic rotations.

The key argument for solving this game is based on the following fact (that was first proved by Leonhard Euler more than 200 years ago):

Any rotation of the globe can be described by a single basic rotation around a single axis.

This holds, no matter how many basic rotations we concatenate in the movement. We can now describe the difference between the placement of the right globe, relative to the left one, by a basic rotation. This basic rotation yields an axis of rotation. Taking the intersection of that axis with the globe yields the 2 points that have the same position on both globes. We are next going to prove the above observation using linear algebra. We believe that if you have not heard about linear algebra before, this chapter might be challenging to follow at first reading. If this is the case, we suggest that you do what we usually do in such cases: Put the book aside for a few days and give it another try later.

7.3 SOME MATHEMATICS: LINEAR ALGEBRA

We argued above that a rotation R can be described as a map that maps any point v on the globe to another point $R(v)$ on the globe; for example, $R(\text{Zurich}) = \text{Copenhagen}$. Here and below, v

is always a 3-dimensional vector, so we write

$$v = \begin{pmatrix} v_1 \\ v_2 \\ v_3 \end{pmatrix}.$$

A few examples of such vectors can be found on page 113. We will see below that our maps R take a very special form. In fact, they can always be written as

$$R\begin{pmatrix} v_1 \\ v_2 \\ v_3 \end{pmatrix} = \begin{pmatrix} a_1v_1 + a_2v_2 + a_3v_3 \\ b_1v_1 + b_2v_2 + b_3v_3 \\ c_1v_1 + c_2v_2 + c_3v_3 \end{pmatrix}, \tag{7.1}$$

for some choice of real numbers a_1, a_2, a_3, b_1, b_2, b_3, and c_1, c_2, c_3. If we turn the globe upside-down around the y-axis, for example, so R(North Pole) = South Pole, then we have

$$R\begin{pmatrix} v_1 \\ v_2 \\ v_3 \end{pmatrix} = \begin{pmatrix} -v_1 \\ v_2 \\ -v_3 \end{pmatrix},$$

which, for suitable constants $a_1, a_2, a_3, b_1, \ldots$, has the form of equation (7.1). Can you find a point v on the globe with $R(v) = v$?

Maps of this form are sometimes called "linear maps" hence the expression "linear algebra". And the question whether there exists a v such that

$$\begin{pmatrix} a_1v_1 + a_2v_2 + a_3v_3 \\ b_1v_1 + b_2v_2 + b_3v_3 \\ c_1v_1 + c_2v_2 + c_3v_3 \end{pmatrix} = 1 \cdot \begin{pmatrix} v_1 \\ v_2 \\ v_3 \end{pmatrix} \tag{7.2}$$

is an instance of a well-studied question: the so-called "eigenvalue" problem. The rotation R is characterized by the constants $a_1, a_2, a_3, \ldots$. If there is a v that satisfies equation (7.2), we say that 1 is an *eigenvalue* of R, and v is an *eigenvector* with eigenvalue 1.

We now argue that there has to be an eigenvalue of 1. Following this part requires some background knowledge in linear algebra. We do our best to explain all the relevant terms, but following some of the arguments might be difficult, and you can also choose to skip to section 7.4.

The Eigenvalue Problem

We first simplify the notation. Equation (7.1) can be written as a matrix-vector multiplication (see appendix B.7). More precisely, defining the matrix

$$R := \begin{pmatrix} a_1 & a_2 & a_3 \\ b_1 & b_2 & b_3 \\ c_1 & c_2 & c_3 \end{pmatrix},$$

then any point v maps to Rv, where Rv is a matrix-vector multiplication. That is, we use the same letter R for the map R and for the matrix R.

So far, we have considered a map R that maps any point on the globe to another point on the globe. It will, however, be easier if we consider the map on the whole 3-dimensional space, that is, on $\mathbb{R}^3$. In fact, we even consider a larger space, namely, $\mathbb{C}^3$. The real numbers are a subset of a larger collection of numbers, the "complex numbers" $\mathbb{C}$, which we introduce in appendix B.8. In short, the space of complex numbers contains even (weird) numbers, such as $\sqrt{-1}$. As for $\mathbb{R}^3$, the set $\mathbb{C}^3$ contains vectors of the form

$$v = \begin{pmatrix} v_1 \\ v_2 \\ v_3 \end{pmatrix},$$

but now each of the components does not have to be real, but can be complex, that is, $v_1, v_2, v_3 \in \mathbb{C}$. We have $\mathbb{R} \subset \mathbb{C}$, and $\mathbb{R}^3 \subset \mathbb{C}^3$.

We call $\lambda \in \mathbb{C}$ an *eigenvalue* of R if there exists a vector $v \in \mathbb{C}^3$ such that

$$Rv = \lambda v,$$

that is, R leaves the vector v unchanged except for multiplication with the eigenvalue. Such a vector v is called an *eigenvector* of R with eigenvalue λ. As we argued above, it suffices to show that the matrix R that describes the rotation of the globe has an eigenvalue $\lambda = 1$. Then, there will be a real[2] eigenvector $v \in \mathbb{R}^3$ such that $Rv = v$, which means that the vector (place on the globe) v is unchanged under the movement R. If v is an eigenvector with eigenvalue 1, then so is $-v$, which means that the location $-v$ on the opposite side of the globe will then also remain unchanged in its position.

Below, we will establish that

(a) there are 3 (not necessarily distinct) eigenvalues of R, whose product equals 1,

(b) all 3 eigenvalues of R have absolute value 1, and

(c) the existence of any complex eigenvalue implies that its complex conjugate is an eigenvalue, too.

These three properties imply that 1 is an eigenvalue of R.

Basic Rotations

Let us first consider basic rotations (i.e., rotations around a single axis), which have a particularly easy form. A basic rotation by an angle $\phi \in [0, 2\pi]$ around the z-axis, for example, will transport the vector v on the sphere to a vector $R_{z,\phi}v$, where $R_{z,\phi}$ is the matrix

$$R_{z,\phi} = \begin{pmatrix} \cos(\phi) & -\sin(\phi) & 0 \\ \sin(\phi) & \cos(\phi) & 0 \\ 0 & 0 & 1 \end{pmatrix}.$$

It leaves the value v_3 along the z-axis unchanged and rotates the coordinates (v_1, v_2) in the x-y-plane. Now, any basic rotation can

2. If $Rv = v$ for a complex, nonreal v, then $Rv = v$ still holds if you set the imaginary parts of v to 0.

be written in the more general form

$$R_{A,\phi} = A \begin{pmatrix} \cos(\phi) & -\sin(\phi) & 0 \\ \sin(\phi) & \cos(\phi) & 0 \\ 0 & 0 & 1 \end{pmatrix} A^\top, \tag{7.3}$$

where $A \in \mathbb{R}^{3\times3}$ with $A^\top A = AA^\top = \mathrm{Id}$ is a matrix that defines a basis transformation ($A^\top$ is the transpose of A and Id is the identity matrix, which contains ones on the diagonal and zeros everywhere else). After the transformation by $A^\top$, the rotation is then just a rotation around the new z-axis by angle ϕ, and the final multiplication with A rotates back into the original coordinate system.

There are two important observations about a basic rotation, that is, a rotation that can be written in the form of equation (7.3):

1. First, the rotation is orthogonal in the sense that for all A and ϕ,

$$R_{A,\phi}^\top R_{A,\phi} = \mathrm{Id},$$

   as $R_{A,\phi}^\top R_{A,\phi}$ equals

$$A \begin{pmatrix} \cos^2(\phi) + \sin^2(\phi) & 0 & 0 \\ 0 & \cos^2(\phi) + \sin^2(\phi) & 0 \\ 0 & 0 & 1 \end{pmatrix} A^\top = AA^\top.$$

2. Second, the determinant of $R_{A,\phi}$ equals 1 (i.e., $\det(R_{A,\phi}) = 1$). This follows because the determinant is multiplicative and hence

$$\begin{aligned} \det(R_{A,\phi}) &= \det(AR_{z,\phi}A^\top) = \det(A)\det(R_{z,\phi})\det(A^\top) \\ &= 1 \cdot \det(R_{z,\phi}) \cdot 1 = \cos^2(\phi) + \sin^2(\phi) = 1. \end{aligned}$$

Arbitrary Rotations

The same two properties now carry over if we concatenate several rotations. Consider the rotations $R_{A,\phi}$, $R_{B,\psi}$, and their product $R := R_{A,\phi} R_{B,\psi}$. The properties described above still hold:

$$R^\top R = R_{B,\psi}^\top R_{A,\phi}^\top R_{A,\phi} R_{B,\psi} = \mathrm{Id}, \tag{7.4}$$

and

$$\det(R) = \det(R_{A,\phi})\det(R_{B,\psi}) = 1. \tag{7.5}$$

By induction, the same holds, of course, for a product of arbitrarily many rotations.

As the determinant equals the product of all eigenvalues, this establishes property (a). Furthermore, eigenvalues of matrices with property (7.4) necessarily have an absolute value of 1:

$$\|v\|^2 = v^\top v = v^\top R^\top R v = \|Rv\|^2 = \|\lambda v\|^2 = |\lambda|^2 \|v\|^2,$$

implying that $|\lambda| = 1$, and therefore establishing property (b). Finally, we are now ready to argue property (c): that complex eigenvalues always come in pairs. One typically writes $\overline{z}$ for the complex conjugate of z, where the imaginary part switches sign. If $z = a + ib$ for $a, b \in \mathbb{R}$, then $\overline{z} = a - ib$. Now, if R has an eigenvalue $\lambda \in \mathbb{C}$ with eigenvector $v \in \mathbb{C}^3$, then $\overline{\lambda}$ is an eigenvalue of $\overline{v}$:

$$R\overline{v} = \overline{R}\overline{v} = \overline{Rv} = \overline{\lambda v} = \overline{\lambda}\overline{v}.$$

We are now able to put together these observations. Consider the three eigenvalues λ_1, λ_2, and λ_3, which are not necessarily distinct. We know that not only their absolute values but also their product equals 1. If two of them are not real, but complex, then the third eigenvalue must be 1 (this is because $\lambda\overline{\lambda} = |\lambda|^2 = 1$). The same follows if all eigenvalues are real. This proves that one of the eigenvalues equals 1.

All matrices $R \in \mathbb{R}^{3\times 3}$ with $R^\top R = \mathrm{Id}$ and $\det(R) = 1$ form a group that is often denoted by $SO(3)$, called the *special orthogonal group*. They all correspond to a basic rotation for a suitable basis transformation A and a rotation angle ϕ. The rotation angle can be easily inferred. The trace of a matrix, trace(R), is the sum of the diagonal entries and is invariant under permuting products, which here implies that $\mathrm{trace}(R) = \mathrm{trace}(AR_{z,\phi}A^\top) =$

$\mathrm{trace}(A^\top A R_{z,\phi}) = \mathrm{trace}(R_{z,\phi})$. The trace of $R_{z,\phi}$ equals $1 + 2\cos(\phi)$, and the sum of the diagonal entries of R hence determines the rotation angle.

7.4 SHORT HISTORY

Many textbooks on linear algebra cover the necessary material [e.g., Fischer, 2002], and they usually use arguments from linear algebra. Instead of considering two globes, one sometimes considers a soccer ball during a soccer match at the beginning of the first half and at the beginning of the second half. The mathematical statement itself was first proved by Leonhard Euler using elementary geometry [Euler, 1776].

7.5 PRACTICAL ADVICE

It is, of course, impossible to describe the location of the invariant points with very high precision. Instead, the players can be asked to name a country after 5 seconds. Then one checks which country is closest to 1 of the 2 invariant points, looking for the closest distance between the invariant point and any point in the country (the invariant point might lie in an ocean, in which case one would try to name a country that is as close as possible). Using countries (instead of cities) also adds an interesting aspect to the game since, say, Russia is more likely to be right a priori than Monaco. So players who do not see right away where the invariant point is can still win by choosing a country with a large area.

The game can in principle be played with objects that are not round, such as a cello. The above arguments still hold: Any sequence of rotations can be described by a single rotation with a single axis of rotation. On this axis all points are invariant. That is, if we describe the right cello as the rotation of the left cello, the intersection of the axis of rotation and the cello yields two points that have the same position on both cellos. Using two spheres, however, has the advantage that there is a natural center that the laser beams point away from (in general, one needs to agree on such a center). We found that in general, nonspherical objects make the game more difficult.

Rather than using globes, the game is perhaps easier to play with two printouts of random rotations of the globe, as seen from a large distance. The figure on page 110 is an example of two such printouts, with Seattle being the stationary point. Of course, this way we might miss places that are on the "backside," that is, on the part of the globe that is not visible. However, this is not really a restriction, since one of the two points of interest will always be visible.

Since the two solutions on the globe will be antipodes (that is, directly opposite on the globe), then one of these two points will be visible to the observer (the one on whichever half of the globe we are looking at). We will thus have exactly one point that is common on the visible part of the globe (assuming that not *all* points are in the same place and that we can really see half of the globe, which is only approximately true if we take a snapshot from space). In the example, Seattle was in a common location on both globes. The antipode is southwest of South Africa, close to Île de la Possession. This second fixed location will become visible as soon as Seattle is rotated out of the visible part of the globe. With the same reasoning, we can consider the following slight variation of the game: One takes a picture from the globe, rotates it, takes another picture, and then asks the audience to spot the place on the globe (e.g., the city) that has the same position in both pictures. To verify, one can then overlay the two pictures. This variation of the game requires that the position in space from which one takes the pictures is fixed.

8 THE FALLEN PICTURE AND ALGEBRAIC TOPOLOGY

Number of players:	1 or more
You will need:	a picture with a long string attached to it; some nails in the wall on which the picture can be hung using the string; it should be possible to remove the nails easily (for an alternative setup: a wooden structure with removable bars; rope)

8.1 THE FALLEN PICTURE

Suppose your friend has changed careers and has decided to become a painter. Since you had advised her to follow her interests, she proudly announces that you are receiving her first completed artwork. She considers her debut work a big success and wants to ensure that it hangs safely on the wall. She therefore watches you hammering 5 nails into the wall that will secure the painting. You attach a string to the picture's frame (top left and top right) that is supposed to wind around the nails. Unfortunately, you do not share your friend's enthusiasm about the painting. You would not be upset if the painting fell down and broke, but since you do not want to hurt your friend's feelings, you do not say anything.

Is there a way to wind the string around the nails, such that the painting hangs on the wall but falls down as soon as one of the nails is pulled out of the wall?

You are looking for a solution that ensures the painting falls down no matter which nail is pulled out.

It is easiest to start with the challenge for 2 nails. The picture should hang on the wall and fall down if either the left or the right nail is pulled out. In most households, pictures are hung by putting the string over two nails at the same time. This standard solution is clearly insufficient for our case: If we pull out the left nail, for example, the picture may lose its horizontal alignment, but it will still hang on the wall due to the right nail. Thus, this configuration cannot be the solution we are looking for—the requirement was that the picture falls down as soon as either one of the 2 nails is pulled out.

What happens if we, instead, hang the picture on the left nail and ignore the right one? The picture would then fall down if we remove the left nail; but it would not if we remove the right nail. Thus, again, this is not the correct solution. The trick must therefore be to wind the string around the 2 nails in a sufficiently complicated manner. We encourage you to take a piece of paper and try to find the solution for 2 nails by trial and error. (There is a solution!)

If you have found the solution, you can take some time to think about the problem a bit further: Is there a way to modify the solution such that it still works? What does the problem have to do with mathematics? And what about more than 2 nails?

8.2 SOLUTION FOR 2 NAILS

The figure on the next page reveals a solution for 2 nails. Convince yourself that the solution works: If we pull out the left nail, the picture will fall down. If we leave the left nail in place and pull out the right nail, the picture also falls down.

Is there a more systematic way to come up with a solution than trial and error? And is there a solution if we have more than 2 nails (e.g., 5, 6, or even 411950)? We first rewrite the game as a problem of lining up dancers at dancing school. We then argue that both problems have the same structure, that is, solving one is as good as solving the other. Perhaps surprisingly, discovering a systematic way of constructing solutions is easier for the dancing problem than for the picture-hanging problem.

8.3 DANCING

Consider the following situation at a dancing school. Two dancing classes (Tango Argentino and Viennese Waltz) are about to start, and all dancers are assembled in one of the classrooms. They are standing in a line, the ones dancing the male part ("males"), those taking the female part ("females"), tango dancers, and waltz dancers mixed. The instructor wants to give some general comments about dancing, but he faces the following difficulty: The dancers are highly motivated. When they are standing next to a possible dancing partner, they immediately move away from the line, start to dance, and stop listening. The remaining dancers close the gaps. New pairs might form and they, too, would start dancing until there are no more matches between immediate neighbors. In this problem, we assume that people dance only with a person who prefers the same dance and who takes the other part (i.e., tango males with tango females, waltz females with waltz males, etc.). For example, the line

(tango-female, waltz-male, waltz-female,
tango-male, tango-female, tango-male)

would dissolve: First, the inner two pairs vanish (waltz-male and waltz-female, and tango-male and tango-female), and then the two dancers on the outside (tango-female and tango-male) become neighbors and start dancing, too. The instructor is now trying to order the dancers such that the following two constraints are satisfied:

1. There is no matching pair in the group, and therefore nobody starts dancing, listening instead to the instructor.
2. As soon as one of the two groups is called out of the room (either the tango dancers or the waltz dancers), the remaining people in the line start dancing, because now (or

after a short time) they find themselves next to a suitable partner.

Suppose there is a group of 4 people: a tango-male, a tango-female, a waltz-male, and a waltz-female. We challenge you to stop reading for a moment, grab a piece of paper, and find a solution for the instructor. The key idea here is to separate the waltz from the tango couple. One possibility is to line them up in the following order:

$$\text{(tango-male, waltz-female, tango-female, waltz-male).} \qquad (8.1)$$

There is no matching pair at the moment, but if the waltz couple leaves the room, the tango people start dancing, and, similarly, if the tango couple leaves, the waltz dance can begin.

How does this dancing problem relate to the falling picture? These two problems are the same! We simply identify tango with the left nail, waltz with the right nail, men with clockwise circles, and women with counterclockwise circles. The line up of dancers described in equation (8.1) corresponds exactly to the solution shown on page 126.

The dancing problem captures the essential structure of the problem. We will see that this is the so-called *topology* of the problem. In section 8.4, we summarize some concepts of algebraic topology. We tried to write the section in a way that is easy to follow, even if you have never encountered topology before. If some parts are not clear after the first reading, try putting the book aside and giving it another try later—for us, this often helps.

8.4 SOME MATHEMATICS: ALGEBRAIC TOPOLOGY

When we study properties of mathematical objects, such as sets, numbers, or groups, it is often helpful to state which properties we are *not* interested in. Objects that differ only in those neglected properties are then considered to be equivalent. In chapter 4, for example, we did not want to distinguish between (the sums) 2 and 5, since they corresponded to the same box. Therefore, we simply defined these numbers to be the same. Formally, we constructed the set $\mathbb{Z}/3\mathbb{Z}$.

In *topology*, we do not distinguish between two sets that can be transformed into each other by stretching, pulling, or poking. A topologist does not distinguish, for example, between two footballs that differ only in the amount of air inside. Similarly, two rubber bands still have the same topological properties, even if one of them is extended to twice its unstretched length. Cutting, however, usually changes the topological properties: If one of the rubber bands rips apart, it becomes topologically different from a nonbroken version. Mathematically, nonripping transformations are usually described by continuous maps (see appendix C.5 for details).

In *algebraic topology*, we can describe an object by a group (see section 4.3). We will first discuss the fundamental group of a set X. This group is denoted by $\Pi_1(X)$ and captures important (topological) properties of the set X. A flat and an inflated football, for example, is each assigned to the same fundamental group. The torn and nontorn rubber bands, on the other hand, are not. The concept of fundamental groups will help us find solutions to the picture-hanging problem for an arbitrary number of nails. We try to explain all the steps. For those of you who have a bit more mathematical training, we provide more precise details in appendix C.5.

Fundamental Group

Consider a space X and a point x_0 that lies in this set. For now, it is easiest to think about X as being a subset of $\mathbb{R}^2$, such as the shaded area in the following figure. Since you can "move around" in X, we decided to call it a space rather than a set.

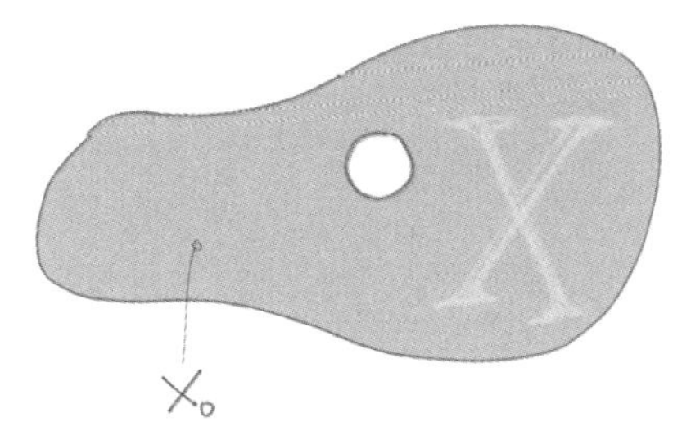

The basic building blocks for the fundamental group of X are loops. These are "walks" in X that start at x_0 and end at x_0. Mathematically, they are continuous maps from $[0, 1]$ into X. As topologists, we do not want to distinguish between loops that are just slightly stretched version of each other. We therefore call all these loops equivalent, or *homotopic*, and regard them as the same thing. This is similar to what we saw in chapter 4, where we identified the sums 2, 5, and 8. We believe that the following figure may help clarify the idea. All three loops in the figure are homotopic to one another. You can think of them as rubber bands that you can stretch.

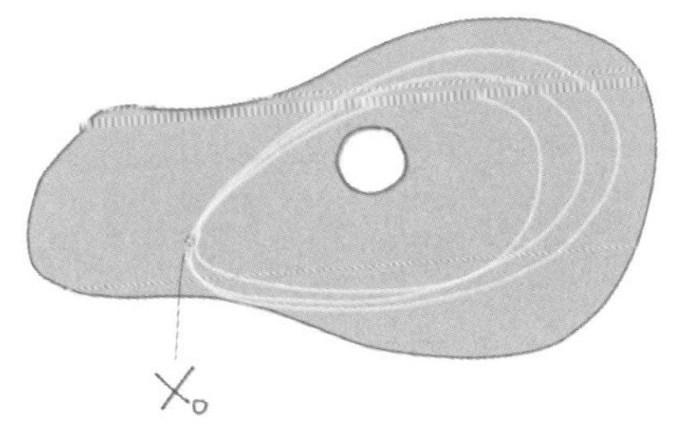

Note that all of them loop around the hole in the middle. It may be helpful to think about the hole as being the trunk of a very large tree growing out of this page. The tree is an obstacle when transforming the rubber bands. Without ripping them apart, the rubber bands will always make exactly one loop around the tree. With the equivalence, or *homotopy*, of loops in mind, when speaking about loops, we will really mean the loop's equivalence class (that is, the loop together with all of its equivalent partner loops).

Loops seem very different from integers, say, but surprisingly, we can still perform computations on loops that are similar to adding numbers. As an example, consider two (different) loops ℓ and m. Similarly to taking two numbers 4 and 5, and computing its sum $4+5$, we can construct a new loop from the pair ℓ and m: the product loop $\ell \circ m$. This new loop is described by first running on the path ℓ, and then around the loop m (we have to run the whole path in double speed, though). This new loop $\ell \circ m$ also starts and ends at x_0. We call $\circ$ an *operation* on the set of loops. In fact, the set of loops with operation $\circ$ forms a *group*. The title "group" can be awarded, because the following four properties hold. Do they ring a bell?

1. The product loop $\ell \circ m$ is again a loop.
2. The loop action is *associative*: For all loops k, ℓ, and m, we have

$$k \circ (\ell \circ m) = (k \circ \ell) \circ m.$$

   (We will not prove this property, but, instead, we ask you to trust us that this is indeed the case.)
3. There exists a neutral element e, such that for all loops ℓ, we have $\ell \circ e = \ell$ and $e \circ \ell = \ell$. These properties are satisfied by a not very exciting loop e: the one that starts in x_0 and ... stays there. Let us call this the *neutral loop*.
4. Finally, for each loop ℓ, there exists another loop, usually called the *inverse loop* ℓ^{-1}, that satisfies $\ell^{-1} \circ \ell = \ell \circ \ell^{-1} = e$. Indeed, for a given loop ℓ, the inverse loop can be defined by running ℓ backward. The concatenated walk is then topologically equivalent (homotopic) to having stayed at x_0 in the first place—even though doing nothing may be less exhausting.

Done—we have a group structure! So far, all loops we have considered start and finish in x_0, so it seems as if the group seems to depend on x_0. For many spaces X, however, the starting point does not have any influence on the group. In particular, this is the case for "path-connected" spaces, that is, spaces in which any 2 points can be connected by a (continuous) path. We can therefore define the *fundamental group* of X as the group of homotopy classes of loops around an arbitrarily chosen starting point $x_0 \in X$. The last sentence sounds complicated, but hopefully we clarified what it means. Topologists usually denote the fundamental group by $\Pi_1(X)$.

So far, the fundamental group appears abstract and not very practical to work with. We will show that for many spaces X, the group actually takes an easy form.

Fundamental Group of a One-Tree Meadow

How can we describe the fundamental group of the example on page 129—the space, let us call it W, with one tree in the middle? Not surprisingly, the tree plays a crucial role here. Any loop that does not circuit the tree is equivalent to the neutral loop e: We can make the loop smaller and smaller until we obtain a loop that barely leaves our starting point x_0. But a loop that circuits the tree (say, clockwise, and exactly one time) is different: There is no way to transform it to the neutral loop without ripping it into pieces. Let us call such a loop a. The figure on page 129 shows some members of this equivalence class. What about the element $a \circ a$? This circuits the tree twice and is different from both a and e. For notational convenience, we do not write $a \circ a$, but rather a^2. Similarly, the loops $a^3 := a \circ a \circ a$ and $a^4 := a \circ a \circ a \circ a$ circuit the tree 3 and 4 times, respectively, and are also not equivalent to the previous loops. Taking the product of a^3 and a^4 yields a loop that circuits the tree 7 times clockwise: $a^3 \circ a^4 = a^7$. What about inverse elements? We can "neutralize" the loop a^7, for example, by taking the product with a loop that walks around the tree seven times in the opposite direction (that is, counterclockwise). We can, of course, denote this new loop however we like, but it is very convenient to call it a^{-7}. The minus in the exponent says that the circuits are performed counterclockwise. Finally, what is the result if we concatenate 7 counterclockwise circuits with 9 clockwise circuits? Exactly 2 clockwise circuits. Summarizing this paragraph, we have the equations

$$a^4 = a \circ a \circ a \circ a, \quad a^3 \circ a^4 = a^7, \quad (a^7)^{-1} = a^{-7}, \quad (a^7)^{-1} \circ a^9 = a^2. \tag{8.2}$$

These computations should remind us of a group that we have seen before: the integers $\mathbb{Z}$ with the addition $+$, but clearly, the notation is somewhat different. It turns out that there exists a one-to-one correspondence between our fundamental group of the one-tree meadow and $\mathbb{Z}$. We say that they are "isomorphic" as groups and write

$$\Pi_1(W) \cong \mathbb{Z}. \tag{8.3}$$

For example, the equations in $\mathbb{Z}$ that correspond to the ones in equation (8.2) read as follows:

$$4 = 1 + 1 + 1 + 1, \quad 3 + 4 = 7, \quad -(+7) = -7, \quad -7 + 9 = 2.$$

A formal proof of equation (8.3) can be found in a book by Allen Hatcher, for example (see section 8.6). It is a famous result in algebraic topology and can be used as a building block for computing the fundamental group of more complex spaces, too. The following theorem, for example, shows us how to compute the fundamental group when connecting two meadows with a single tree each to a large meadow with two trees.

The Seifert—van Kampen Theorem and a Two-Tree Meadow

Different operations allow us to combine several spaces into more complex structures. One of them is called the "gluing operation," usually referred to as the wedge sum. One takes two spaces X and Y and glues them together by a small bridge. The example of the space $W \vee W$ (W "wedge" W), obtained by taking the wedge sum of the one-tree meadow W with itself, is shown in the figure below and should suffice to understand the concept.

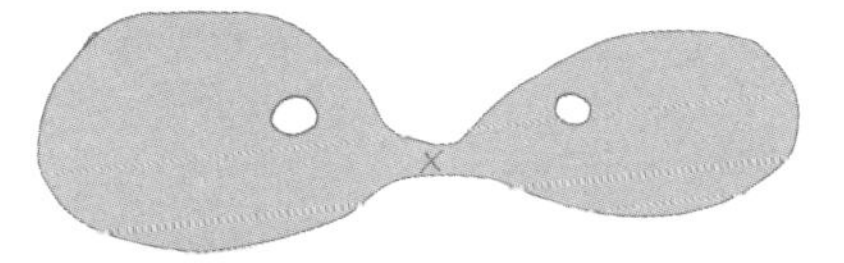

As a topologist, we would ask immediately, of course, what the fundamental group of the new structure looks like. The following theorem by van Kampen and Seifert states that it is straightforward to compute the fundamental group of the wedge sum if the individual fundamental groups are known. The result involves the free product of groups, a concept that we explain below. Let us first state the result. Consider spaces $X_1, X_2, \ldots, X_n$. Then the fundamental group of the wedge sum can be computed from the individual fundamental groups:

$$\Pi_1(X_1 \vee X_2 \vee \cdots \vee X_n) \cong \Pi_1(X_1) * \Pi_1(X_2) * \cdots * \Pi_1(X_n). \quad (8.4)$$

Here, $*$ denotes the free product of groups. As a direct consequence, we have that

$$\Pi_1(W \vee W) \cong \mathbb{Z} * \mathbb{Z}. \quad (8.5)$$

So, how is the free product of two groups defined? The first factor in $\mathbb{Z} * \mathbb{Z}$ corresponds to the fundamental group connected to the first tree. Recall that we have written a^4 instead of 4. Following this convention, let us write b^4 instead of 4 for the second factor in $\mathbb{Z} * \mathbb{Z}$. (The loop b^4 will correspond to making

---

Solution to the figures at the end of section 3.4. The correct cards are the ten of spades (first example) and ten of clubs (second example). The score in the latter example is 57.

4 clockwise circuits around the second tree.) The free product $\mathbb{Z} * \mathbb{Z}$ now contains "words" that concatenate elements of both groups, such as

$$ab^{-4}b^3b^2a^{-1}a^3b^{-1}ba. \tag{8.6}$$

It was a good idea to write a^4 and b^4 instead of 4, since otherwise this expression would read $1-4+3+2-1+3-1+1+1$ and we would not be able to see which part of the word comes from the first and which part from the second version of $\mathbb{Z}$. The word (8.6) is an element of the fundamental group $\Pi_1(W \vee W)$, and as such, it must correspond to a loop. The loop is constructed by circuiting the left tree once (clockwise), then circuiting the right tree four times (counterclockwise), then three times the right tree (clockwise), and so forth. This operation can be simplified to

$$ab^{-4}b^3b^2a^{-1}a^3b^{-1}ba = a(b^{-4}b^3b^2)(a^{-1}a^3)(b^{-1}b)a = aba^3.$$

As before, the group action is the concatenation of such words:

$$b^5aba^2 \circ a^{-1}b^{-6} = b^5abab^{-6}.$$

Certainly, you can find the inverse element of this word.[1] We have learned to perform computations using loops in relatively complex spaces. This allows us to solve the original problem for 2 nails in a systematic way, which can easily be extended to more than 2 nails.

8.5 SOLUTION, CONTINUED

The key observation is that the wall corresponds to the space $W \vee W$, with the nails taking the role of the trees. The string winding around the nails represents a homotopy class of loops.

---

1. The inverse element of b^5abab^{-6} is $b^6a^{-1}b^{-1}a^{-1}b^{-5}$ as their product equals $b^5abab^{-6} \circ b^6a^{-1}b^{-1}a^{-1}b^{-5} = b^5aba(b^{-6}b^6)a^{-1}b^{-1}a^{-1}b^{-5} = \cdots = e$.

That is, each candidate solution corresponds to an element in the fundamental group $\Pi_1(W \vee W)$. We now need to translate the problem description of the picture-hanging problem into properties of the solution element in $\Pi_1(W \vee W)$. The picture hangs on the wall if and only if the element is not equivalent to the neutral loop. But how can we translate the property that removing any of the nails results in a falling picture? To do so, we need to understand how to mathematically describe the removal of a nail. Consider the word $ab^2a^{-2}b^3a^2$, for example. Removing the second nail lets all the loops around the second nail collapse—the second tree in the meadow is gone. Mathematically, this corresponds to removing all b components from the word, so that $ab^2a^{-2}b^3a^2$ becomes $aa^{-2}a^2 = a$. The key question therefore becomes:

> *Is there an element in* $\Pi_1(\mathrm{W} \vee \mathrm{W})$*, different from the neutral loop, such that removing either all* a *components or all* b *components yields the neutral loop* e*?*

The answer is yes. It is achieved by the word

$$s(a, b) := ab^{-1}a^{-1}b, \tag{8.7}$$

for example. Please try to draw this solution. It should yield the same figure as the one on page 126. (Hint: Recall that a represents a clockwise wind around the left nail; also, your drawing becomes easier to interpret if you make sure that there are no more than 2 lines that cross at a single point.)

Equation (8.7) shows the solution for 2 nails. Surprisingly, we can take this as a starting point to construct a solution for an arbitrary number of nails. Can you figure out how? The proof goes by induction. Suppose that we have a solution s (that is, a word) for $n \geq 2$ nails. We can then construct a solution for $n+1$ nails by $szs^{-1}z^{-1}$, where z denotes the loop around the new $(n+1)$th nail. The solutions for 3 and 4 nails become

$$s(a,b,c) := ab^{-1}a^{-1}bcb^{-1}aba^{-1}c^{-1} \tag{8.8}$$

$$s(a,b,c,d) := ab^{-1}a^{-1}bcb^{-1}aba^{-1}c^{-1}dcab^{-1}a^{-1}bc^{-1}b^{-1}aba^{-1}d^{-1}. \tag{8.9}$$

We challenge you to draw these solutions for the falling-picture problem. A representation for 3 different dance styles (indicated by colors) is shown below:

If you remove, for example, the blue dancers that are dancing a swing, then the remaining dancers will also naturally form pairs and the whole line will dissolve. The outcome is identical if you remove the green tango dancers or the red waltz dancers.

Solution Length

The length of the solution presented earlier grows exponentially with the number of nails: An induction argument shows that the length of the above solution for n nails equals

$$2^n + 2^{n-1} - 2.$$

These solutions very quickly become impractical. Researchers have therefore tried to find shorter solutions to the problem.

Some solutions grow "only" polynomially with the number of nails. The following table shows what difference this makes in practice. We can save a lot of string!

number of nails	exponential solution length	polynomial solution length
1	1	1
2	4	4
3	10	10
4	22	16
5	46	28
6	94	40
7	190	52
8	382	64
9	766	88
10	1534	112
⋮	⋮	⋮
50	1688849860263934	2752

8.6 SHORT HISTORY

Equation (8.3) can be found as theorem 1.7 in Hatcher [2002], for example. The Seifert–van Kampen theorem dates back to Van Kampen [1933] and Seifert [1931]. The solution that grows only polynomially in the number of nails is from Pegg [2002] and is presented in Demaine et al. [2012], for example. The prisoner idea (see section 8.7) we developed jointly with Anders Tolver.

8.7 PRACTICAL ADVICE

Before playing this game, you should practice drawing the solutions a couple of times. In particular, intersections of different lines can be hard to interpret. We therefore suggest always avoiding intersections of more than 2 lines.

The inductive construction also provides a hint for drawing the solutions. Start with a solution for 2 nails, say, add a new nail with a clockwise circuit, follow the original solution backward, and then add one counterclockwise circuit around the new nail. This solution, however, is not always the easiest when doing a visual check.

When dealing with 4 or 5 nails, you might want to consider using the polynomial solution rather than the exponential one.

There is one more subtle aspect that you need to keep in mind: The solution may not work anymore if you produce knots when winding the string around the nails. One possibility to prevent this from happening is to start with a closed string and to always place the string loops on top of each other.

Alternative Setup

There is an alternative setup of the problem, in which the audience takes a slightly more active role. In that game, some participants are asked to help "arrest" one member of the audience (for some made-up reason). The prison is supposedly high security, and the convicted person has to be tied to 5 bars. The participants are friends of the convicted person and know that the prison hosts one friendly prison guard who usually removes one of the bars, but they do not know in advance which bar will be removed. They therefore want to tie up their friend to the bars such that the friend will be able to escape as soon as one arbitrarily chosen bar is removed. This setup requires a rope and some wooden structure with several bars that can be removed.

APPENDIX A WHAT DO WE MEAN WHEN WE WRITE ...?

If, while reading the book, you encounter some symbol that you do not understand, this table helps explain what we mean.

$\{2,5,6\}$	*a set*: you can imagine this as a bag that contains the numbers 2, 5, and 6.
$5 \in \{2,5,6\}$	*element in a set*: indicates that a certain element (e.g., 5) is contained in the set (e.g., $\{2,5,6\}$).
$\#\{2,5,6\}$	*number of elements in a set*: $\#\{2,5,6\}=3$; also denoted by $\lvert\{2,5,6\}\rvert$.
$\{2,5,6\}\setminus\{2,6\}$	*set difference*: removes certain elements from a set; here: $\{2,5,6\}\setminus\{2,6\}=\{5\}$.
$\{2,5\}\cap\{5,7\}$	*set intersection*: contains all elements that are in both sets; here: $\{2,5\}\cap\{5,7\}=\{5\}$. If the intersection is empty, the sets are called *disjoint*.
$\{2,5,6\}^k$	*power of a set*: this set contains 3^k k-tuples (e.g., $\{2,5,6\}^2=\{(2,2),(2,5),(2,6),(5,2),(5,5),(5,6),(6,2),(6,5),(6,6)\}$).
$\mathbb{N}$	*natural numbers*: $0,1,2,3,\ldots$.
$\mathbb{Z}$	*integers*: as above, but includes negative numbers, too: $\ldots,-3,-2,-1,0,1,2,3,\ldots$.

$\mathbb{Q}$	*rational numbers*: all fractions, such as $2/3$, $-4/17$, and $122/4$, but also 0 and -32.
$\mathbb{R}$	*real numbers*: all rational numbers, together with so-called "irrational numbers," such as $\sqrt{2}$, $\sqrt[5]{7}$, and $\pi \approx 3.1416$.
$\mathbb{C}$	*complex numbers*: each complex number is of the form $a + i \cdot b$ for some $a, b \in \mathbb{R}$. We can add, subtract, multiply, and divide complex numbers, using $i^2 = -1$ (e.g., $(2 + i \cdot 3) + (1 + i \cdot 5) = 3 + i \cdot 8$ and $(4 + i \cdot 1)^2 = 15 + i \cdot 8$); see appendix B.8.
$\lim_{n\to\infty} a_n$	*limit of a sequence*: see appendix B.2.
$a!$	*factorial of a*: in this book, defined for natural numbers; $a! = a \cdot (a-1) \cdot (a-2) \cdot \dots \cdot 2 \cdot 1$ (e.g., $4! = 4 \cdot 3 \cdot 2 \cdot 1 = 24$ and $14! = 87178291200$). Usually, $0!$ is defined as 1.
a^b	*exponentiation*: if b is a natural number, this is short-hand notation for $a \cdot a \cdot \dots \cdot a$ (b times) (e.g., $2^3 = 2 \cdot 2 \cdot 2 = 8$); see appendix B.3 for the general case.
$a \mapsto f(a)$	*function (or mapping)*: you can think of this as a machine that takes some input a and outputs $f(a)$ (e.g., $f(a) = \frac{1}{a}$). To be more specific, one sometimes add $f : A \to B$ to indicate what type of elements are allowed to enter f, and what type of elements are output. In the example above, one could write $f : \mathbb{R} \setminus \{0\} \to \mathbb{R}$.
exp	*exponential function*: arguably the most important function in mathematics; see appendix B.3.

$\log_2$ *logarithm to the base 2*: related to the inverse function of exp; see appendix B.3.

$\sum_{a \in A} f(a)$ *sum*: a short-hand notation for sums with many summands (e.g., $\sum_{a \in \{2,5,7\}} a = 2 + 5 + 7$, and $\sum_{a=13}^{20} \frac{1}{a} = \frac{1}{13} + \frac{1}{14} + \cdots + \frac{1}{20}$).

P *probability*: the probability of a certain event; see appendixes B.5 and B.6.

E *expectation*: the value of a random variable that we expect to see on average. Also called the *mean* of that variable; see appendixes B.5 and B.6.

APPENDIX B WHAT IS ...

B.1 ... A BINARY NUMBER?

We are accustomed to writing numbers using the decimal system. We use the decimal digits 0, 1, 2, 3, 4, 5, 6, 7, 8, and 9 and are comfortable with writing that a year consists of (roughly) 365 days. But why do we write down numbers using 10 digits? It is often speculated that this choice is related to the number of fingers and toes we have, which were used to perform simple calculations ("digitus" is Latin for "finger"). Nowadays, however, most numbers are processed by computers. And electrical circuits are not very well suited to represent the digits 0, 1, ..., 10. Instead, computer systems mostly use the digits 0 and 1 and can represent them by low and high voltage, for example. In a *binary system*, each number is represented using 0s and 1s. Here, the key idea is to use powers of 2. For example, we have

$$11010 = \mathbf{1} \cdot 2^4 + \mathbf{1} \cdot 2^3 + \mathbf{0} \cdot 2^2 + \mathbf{1} \cdot 2^1 + \mathbf{0} \cdot 2^0.$$

In the decimal system, this number is written as 26. For decimal numbers, we perform the same kind of calculations:

$$365 = \mathbf{3} \cdot 10^2 + \mathbf{6} \cdot 10^1 + \mathbf{5} \cdot 10^0,$$

but since we are so used to the system, we do not consciously think about this operation any more.

numeral system	
binary	decimal
0	0
1	1
10	2
11	3
100	4
101	5
110	6
111	7
1000	8
1001	9
1010	10
1011	11
1100	12
⋮	⋮

Sometimes a subscript is used to distinguish between the two systems. For example,

$$110_2 = 6_{10}.$$

It is possible, of course, to use bases other than 2 or 10. The *hexadecimal* system, for example, uses the base 16, and the *octal* system uses 8. The latter also forms the basis of a famous mathematical joke:

> Why do many mathematicians always confuse Halloween and Christmas? Because Oct 31 = Dec 25.

B.2 … A CONVERGING SEQUENCE OR SERIES?

For our purposes, a sequence $(a_n)_n := (a_n)_{n \in \mathbb{N}}$ is an infinite collection of real numbers a_n, where $n = 1, 2, \ldots$ (e.g., 1, $1/2, 1/3, 1/4, 1/5, \ldots$ if $a_n := 1/n$). Even though this sequence never reaches 0, it comes arbitrarily close to it. The sequence defined by $a_n := (-1)^n$, in contrast, does not approach any number, since it jumps between -1 and 1. Formally, one says that a sequence $(a_n)_n$ *converges to a real number* a if for all $\varepsilon > 0$, there exists a natural number N, such that for all $n > N$, we have

$$|a_n - a| < \varepsilon.$$

That is, after some value of N, all elements of the sequence remain in an ε-band around a. We write $a = \lim_{n\to\infty} a_n$.

Some sequences consist of partial sums, which means that each a_n can be written as

$$a_n = \sum_{k=1}^{n} b_k$$

for some sequence $(b_k)_k$. These sequences are often called *series*, and if they converge, their limit is denoted by

$$\lim_{n\to\infty} a_n = \lim_{n\to\infty} \sum_{k=1}^{n} b_k =: \sum_{k=1}^{\infty} b_k.$$

As an example, consider eating a cake by always cutting the remaining piece into two equal-sized halves and eating one of them. You start by having eaten $\frac{1}{2}$ of the cake, then $\frac{1}{2} + \frac{1}{4}$, then $\frac{1}{2} + \frac{1}{4} + \frac{1}{8}$, and so on. Clearly, you never stop eating, and the amount of cake in your stomach is growing, but at the same time, you will never have eaten more than the whole cake. Mathematically, we have

$$\sum_{k=1}^{\infty} \frac{1}{2^k} = \frac{1}{2} + \frac{1}{4} + \frac{1}{8} + \cdots = 1.$$

In high school, one sometimes considers numbers of the form $0.3\bar{9}$, which should really be understood as $0.3 + \sum_{k=2}^{\infty} 9 \cdot 10^{-k} = 0.4$.

B.3 ... AN EXPONENTIAL FUNCTION?

One of the most important series in mathematics defines the *exponential function* exp:

$$\exp(x) := \sum_{k=0}^{\infty} \frac{x^k}{k!}. \tag{B.1}$$

One can show that this series converges for any real number x, that is, $\exp : \mathbb{R} \to \mathbb{R}_{>0}$ is a function from $\mathbb{R}$ to strictly positive real numbers. The value $e := \exp(1) \approx 2.7182$ is the famous Euler's number.

The function exp is strictly monotonically increasing and has an inverse function, the *natural logarithm* $\log : \mathbb{R}_{>0} \to \mathbb{R}$. That is, $\log(a) = b$ if and only if $\exp(b) = a$. The graphs of the functions exp and log are shown in the following figure.

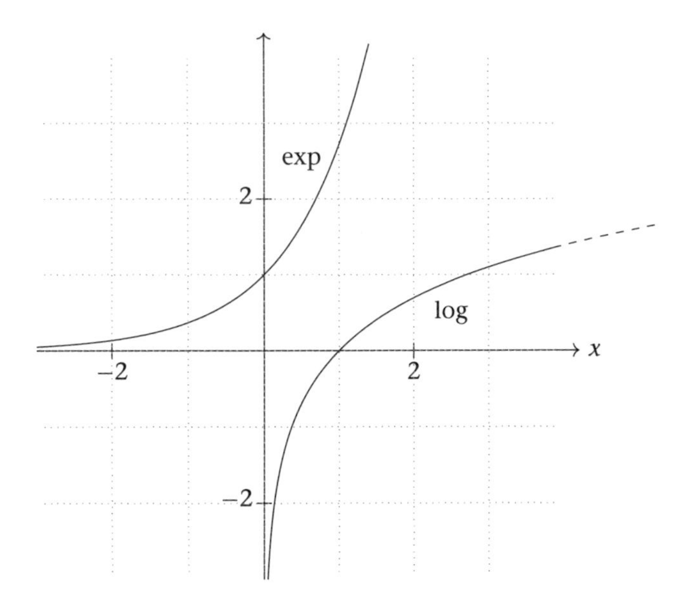

The logarithm function is strictly increasing, but the increase is slow. How slow can be experienced by doing a small experiment.

Turn this book so that the two arrows in the following figure point toward the North Pole and the center of the Earth, respectively. Open the book so that the arrow with the x on page 136 points toward the North Pole, too.

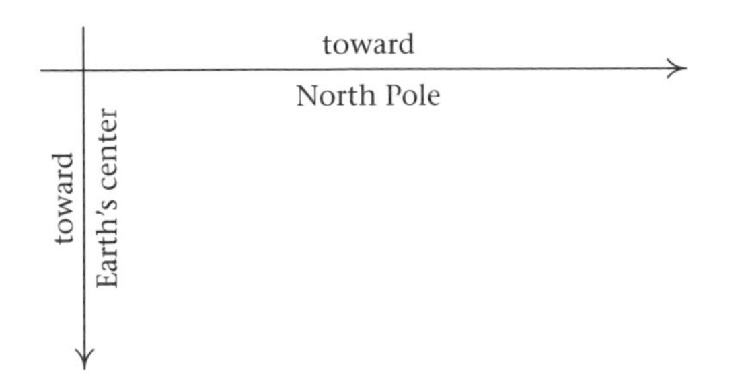

Return now to the previous figure, which shows the two graphs. Assume that you take a pencil and continue to draw the graph of the logarithm (following the dashed line). You draw over this page, leave this book, and continue to hold the pencil at the height of the corresponding value. If you go farther and farther, you will circle the Earth (passing the North Pole) and eventually, after a long time, coming back to page 146 of this book. Where will the graph meet the vertical axis on that page? The Earth's circumference (at the poles) is roughly 40,000 km, and the scale of the drawing is 1 cm. Furthermore,

$$\log(40 \cdot 10^8) \approx 22.1,$$

which means that after circling the Earth, the graph of the logarithm is only about 22 cm above the origin of the coordinate system (that is, the intersection of the two arrows)!

The logarithm and the exponential can be used to extend the exponentiation a^b to cases where the exponent b is not an integer: $a^b := \exp(b \log(a))$. This allows us to define the function $x \mapsto 2^x$, for example. In information theory (chapter 5) and coding theory (chapter 1), one usually considers logarithms to the base 2. This function is the inverse function of $x \mapsto 2^x$ and is often denoted by $\log_2$. We have $\log_2(8) = 3$, because $2^3 = 8$. It holds that

$$\log_2(x) = \frac{\log(x)}{\log(2)}.$$

B.4 ... A BINOMIAL COEFFICIENT?

In the field of combinatorics, we count things. Often, this means counting possibilities. How many towers can I build with 4 different colored toy blocks? In how many ways can we place our 40 guests at 3 tables? How many ways are there to distribute the music parts in a string octet with 4 violins, 2 violas, and 2 cellos, if the musicians do not switch instruments?

The *binomial coefficient* is about counting subsets. Consider the group

$$\{\text{Anna, Bahar, Chane, Dora, Emilian}\},$$

for example. There are 10 possible ways of forming a group of size 3:

$$\begin{array}{cc}
\{\text{Anna, Bahar, Chane}\} & \{\text{Anna, Bahar, Dora}\} \\
\{\text{Anna, Bahar, Emilian}\} & \{\text{Anna, Chane, Dora}\} \\
\{\text{Anna, Chane, Emilian}\} & \{\text{Anna, Dora, Emilian}\} \\
\{\text{Bahar, Chane, Dora}\} & \{\text{Bahar, Chane, Emilian}\} \\
\{\text{Bahar, Dora, Emilian}\} & \{\text{Chane, Dora, Emilian}\}.
\end{array}$$

We can also count the number of combinations without writing them down explicitly. There are 5 possibilities of choosing the first person, 4 possibilities of choosing the second person, and lastly, there are 3 persons left, from which we can choose the last person. The number of combinations we obtain in this way can be written as

$$5 \cdot 4 \cdot 3 = \frac{5 \cdot 4 \cdot 3 \cdot 2 \cdot 1}{2 \cdot 1} = \frac{5!}{(5-3)!}, \tag{B.2}$$

where 5! denotes 5 factorial. Since we consider groups (sets) of people here, we do not want to distinguish between

$$\{\text{Emilian, Anna, Dora}\} \text{ and } \{\text{Dora, Emilian, Anna}\},$$

for example. In equation (B.2), however, both of these combinations have been included. In fact, not only 2 versions of this set are included, but also the other 4 permutations of the set. In total, each set has been counted $6 = 3!$ times. Thus, there are

$$\binom{5}{3} := \frac{5!}{3!(5-3)!} = \frac{120}{12} = 10 \tag{B.3}$$

possibilities to form a set of 3 people out of a set of 5 people.

This derivation may seem a bit overly complex, but it generalizes to sets and subsets of arbitrary size. In general, the *binomial coefficient* is defined as

$$\binom{n}{k} := \frac{n!}{k!(n-k)!}$$

and denotes the number of possibilities of forming subsets of size k out of n distinct elements. The argument for that statement is exactly the same as the one above for $\binom{5}{3}$.

The binomial coefficient becomes important in lottery games, where participants have to guess which 6 out of 49 numbers are drawn, for example. Most of us have relatively little intuition about how many possibilities there are, but when we calculate the numbers, they are usually very large, which, unfortunately, yields very small winning probabilities, for example,

$$\binom{49}{6} = 13983816.$$

The *multinomial coefficient* is an extension of the binomial coefficient. There are

$$\binom{n}{k_1, \ldots, k_m} = \frac{n!}{k_1! \cdots k_m!}$$

possibilities to distribute n numbers into m bins of size $k_1, k_2, \ldots, k_m$ (so $k_1 + \cdots + k_m = n$). For example, given the set of 5 people mentioned above, we want to form one group of 2 people making dinner, one group of 2 people making dessert, and one group of 1 person doing the dishes. There are

$$\binom{5}{2,2,1} = \frac{5!}{2!2!1!} = 30$$

possibilities to form such groups. Also,

$$\binom{n}{k, n-k} = \frac{n!}{k! \cdot (n-k)!} = \binom{n}{k},$$

and we recover the binomial coefficient.

B.5 ... A PROBABILITY?

Probabilities of Events

We will define probabilities for sets (one can think about sets of possible outcomes), which are often called *events*. Consider a (finite) set Ω, which represents the set of possible outcomes. A *probability* is a function P that maps any subset A of Ω to a number between 0 and 1, such that the following three statements hold:

- $P(\Omega) = 1$,
- $P(A^C) = 1 - P(A)$ for all $A \subseteq \Omega$, and
- $P(A \cup B) = P(A) + P(B)$ for all $A, B \subseteq \Omega$ that are disjoint.

Here, $A^C := \Omega \setminus A$ denotes the complement of A, and A and B being disjoint means that they have no element in common: The intersection $A \cap B$ equals the empty set $\emptyset$.

Many random processes can be modeled with such probabilities. The outcome of rolling a die is a famous example and can be modeled by

$$\Omega := \{1, 2, 3, 4, 5, 6\} \text{ and } P(A) := \frac{\#A}{6},$$

where $\#A$ denotes the number of elements in A. Rolling two dice can be modeled using

$$\Omega := \{(1,1), (1,2), (1,3), \ldots, (4,6), (5,6), (6,6)\} \text{ and } P(A) := \frac{\#A}{36}.$$

For example, we have

$$P(\text{sum}=8)=P(\{(2,6),(3,5),(4,4),(5,3),(6,2)\})=5/36$$

because there are 5 pairs (i,j) that yield a sum of 8.

In this book, we are often interested in the probability $P(\text{lose})$ of losing a game, for example. As in the die example, this often requires us to count the number of situations in which we lose and divide that number by the number of possible outcomes.

Independence

We say that two events are *independent* if the probability that both events happen is the product of the probabilities of each event happening. Formally, we say $A, B \subseteq \Omega$ are independent if and only if

$$P(A \cap B) = P(A) \cdot P(B).$$

Intuitively, this means that the occurrence of event A does not tell you anything about the occurrence of event B. Consider, for example, the event A: "sum of the two dice is divisible by 3" and event B: "second die shows a 6." Using the above probability model, this can be written as

$$A := \{(1,2),(1,5),(2,1),(2,4),(3,3),(3,6),(4,2),(4,5),$$
$$(5,1),(5,4),(6,3),(6,6)\} \text{ and}$$

$$B := \{(1,6),(2,6),(3,6),(4,6),(5,6),(6,6)\}.$$

Clearly, we have

$$P(A \cap B) = P(\{(3,6),(6,6)\}) = \frac{2}{36} = \frac{12}{36} \cdot \frac{6}{36} = P(A) \cdot P(B).$$

The events A: "sum of the two dice equals 11" and B: "second die shows a 6," however, are not independent.

B.6 ... AN EXPECTATION?

As in appendix B.5, we start with a (finite) set Ω of possible outcomes and a probability P that maps any subset $A \subseteq \Omega$ to the interval $[0, 1]$. Sometimes, any outcome can be attached to a number that we are particularly interested in. In the example of rolling two dice, for example, we may be particularly interested in the sum of the two dice. This can be achieved by considering the map $X : \Omega \to \mathbb{R}$, with $(i, j) \mapsto i + j$. Such maps X are called *random variables*. Rather than looking at events, we can then compute probabilities involving random variables directly. In the die example, we can write the term $P(\text{sum} = 8)$ as $P(X = 8)$.

Given a set Ω of possible outcomes, a probability P, and a random variable X, we can now compute the *expectation* of X. It is defined as

$$\mathbf{E}X := \sum_x xP(X = x).$$

Here, the sum is over all values that the random variable X can take. For example, for the sum of two dice, we have:

$$\mathbf{E}X = 2 \cdot \frac{1}{36} + 3 \cdot \frac{2}{36} + 4 \cdot \frac{3}{36} + \cdots + 11 \cdot \frac{2}{36} + 12 \cdot \frac{1}{36} = 7.$$

There are two intuitive interpretations of that statement. First, the expectation is the "best guess" for the outcome of X.[1] Second, the convergence to the expectation tells us that when we average a large number of rolls of two dice, the average will be roughly 7 (this is usually referred to as the "law of large numbers").

In chapter 4 on page 65, we considered a random variable for counting mistakes and argued that

$$P(\text{mistakes} = 1) \geq 1 - \frac{1}{p}.$$

1. More formally, we have for all $a \in \mathbb{R}$ that $\mathbf{E}[(X - \mathbf{E}X)^2] \leq \mathbf{E}[(X - a)^2]$, if $\mathbf{E}X^2 < \infty$.

This clearly implies that

$$\mathbb{E}(\text{mistakes}) \geq 1 - \frac{1}{p}.$$

B.7 ... A MATRIX?

A matrix is a collection of numbers organized into rows and columns. For example,

$$A = \begin{pmatrix} 5 & -\pi & 2 \\ 0 & 3 & 3 \end{pmatrix}$$

is called a 2×3 matrix, since it has 2 rows and 3 columns. We often write $A \in \mathbb{R}^{2\times 3}$.

The matrix multiplication of two matrices A and B in general is defined if A has as many columns as B has rows. Then the entry (k, j) of the product AB is calculated by "overlaying" row k of matrix A with the jth column of B and summing up the pairwise entries as

$$(AB)_{k,j} = \sum_{\ell} A_{k,\ell} B_{\ell,j}, \tag{B.4}$$

where the index ℓ in the sum runs over all values of ℓ from 1 to the number of columns of A.

A special but important case of a matrix-matrix multiplication is the matrix-vector multiplication. Writing out equation (B.4), we obtain

$$\begin{pmatrix} a_1 & a_2 & a_3 \\ b_1 & b_2 & b_3 \\ c_1 & c_2 & c_3 \end{pmatrix} \begin{pmatrix} v_1 \\ v_2 \\ v_3 \end{pmatrix} = \begin{pmatrix} a_1v_1 + a_2v_2 + a_3v_3 \\ b_1v_1 + b_2v_2 + b_3v_3 \\ c_1v_1 + c_2v_2 + c_3v_3 \end{pmatrix},$$

which we use in section 7.3, for example.

Many maps that appear in practice have such a form. For example, for a 3×5 matrix A, we can consider the map

$$\begin{array}{ccc} \mathbb{R}^5 & \to & \mathbb{R}^3 \\ v & \mapsto & Av. \end{array}$$

This is often called a *linear map*, because we have the following two properties: first, $A \cdot 0 = 0$, and second, for all $\lambda \in \mathbb{R}$ and v, $w \in \mathbb{R}^5$, we have $A(\lambda v + w) = \lambda Av + Aw$. In general, an $n \times m$ matrix A defines a linear map $\mathbb{R}^m \to \mathbb{R}^n$.

The *transpose* $A^\top$ of a matrix A is obtained by simply turning around rows and columns:

$$A = \begin{pmatrix} 5 & -\pi & 2 \\ 0 & 3 & 3 \end{pmatrix} \quad \Longrightarrow \quad A^\top = \begin{pmatrix} 5 & 0 \\ -\pi & 3 \\ 2 & 3 \end{pmatrix}.$$

B.8 ... A COMPLEX NUMBER?

Complex numbers are, unlike their name suggests, not particularly difficult. One way to think about them is to consider the real numbers $\mathbb{R}$ and to "add" a whole bunch of other numbers. But why would we want to do that? The real numbers have a lot of nice properties, but they are not *complete*. If we consider the following polynomial f,

$$f(x) = x^2 - 2x - 3,$$

then f has two roots: $x = -1$ and $x = 3$. That is, $f(-1) = 0 = f(3)$. In fact, you can write

$$f(x) = (x + 1)(x - 3),$$

which makes it even more obvious what the roots are. But what about a polynomial g with

$$g(x) = x^2 - 2x + 3?$$

We do not find any real number x that satisfies $g(x) = 0$. It somehow feels it should be possible to write

$$g(x) = (x - r)(x - s),$$

but what are r and s? These are complex numbers.

Formally, a complex number $z \in \mathbb{C}$ can be written as

$$z = a + i \cdot b,$$

for $a, b \in \mathbb{R}$. We call a the *real part* and b the *imaginary part* of z. As for real numbers, we can add, substract, multiply, and divide complex numbers, but similarly to when we first learned how to add two ratios, we also need to learn the rules for adding complex numbers. Let

$$z_1 = a_1 + i \cdot b_1,$$
$$z_2 = a_2 + i \cdot b_2.$$

We then define

$$z_1 + z_2 := (a_1 + b_1) + i \cdot (b_1 + b_2),$$
$$z_1 - z_2 := (a_1 - b_1) + i \cdot (b_1 - b_2),$$
$$z_1 \cdot z_2 := (a_1 a_2 - b_1 b_2) + i \cdot (a_1 b_1 + a_2 b_2).$$

Dividing two complex numbers is possible, too, but that is not so important for our book.

The 'i' is really just a symbol, but if you think of it as the famous $\sqrt{-1}$, you may find it easier to memorize the multiplication rule, because you can then simply multiply out and collect terms:

$$\begin{aligned} z_1 \cdot z_2 &= a_1 a_2 + i a_1 b_1 + i a_2 b_2 + i^2 b_1 b_2 \\ &= (a_1 a_2 - b_1 b_2) + i \cdot (a_1 b_1 + a_2 b_2). \end{aligned}$$

What about our polynomial g from above? This polyonomial indeed has two roots. If you like, you can convince yourself that

$$g(1 + i \cdot 2) = 0 \text{ and } g(1 - i \cdot 2) = 0. \tag{B.5}$$

This allows us to write

$$g(x) = (x - 1 - i \cdot 2)(x - 1 + i \cdot 2),$$

which yields the values of r and s we were looking for.

In fact, there is a famous theorem stating that any polynomial g of degree n can be written as

$$g(x) = (x - z_1) \cdot (x - z_2) \cdot \cdots \cdot (x - z_n)$$

with z_is that are not necessarily real but may be complex.

As in equation (B.5), complex numbers often appear in pairs: If

$$z = a + i \cdot b$$

satisfies a certain property, then often $a - i \cdot b$ does, too. We therefore write

$$\bar{z} = a - i \cdot b$$

and call this the *complex conjugate* of z.

APPENDIX C
CHAPTER-SPECIFIC DETAILS

C.1 CHAPTER 1: HAT COLORS AND HAMMING CODES

The construction of Hamming codes for $n = 2^m - 1$, $m \in \mathbb{N}$, with $m > 2$ is maybe easiest when considering linear codes. To do so, we consider codewords such as (010) as an element in $\mathbb{Z}_2^3$ and will take their sums, for example. Here, writing $\mathbb{Z}_2$ is short for $\mathbb{Z}/2\mathbb{Z}$, which works exactly as the cyclic group $\mathbb{Z}/3\mathbb{Z}$ that we discuss in chapter 4. It means that we compute everything modulo 2. For example,

$$(010) + (100) = (110) \quad \text{or} \quad (011) + (111) + (001) = (101).$$

We can also multiply the sequences by 0 or 1:

$$0 \cdot (011) = (000) \quad \text{and} \quad 1 \cdot (011) = (011).$$

This is not a very exciting operation, but it allows us to consider the space $\mathbb{Z}_2^3$ as a so-called "linear space."

More importantly, we call a code W (that is, the collection of codewords) a *linear code* if W is a linear subspace of $\mathbb{Z}_2^n$. Or, equivalently, if the all-zero sequence is in W and if for all $x, y \in W$, we have $x + y \in W$, then W is a linear code. Later in this appendix we will introduce Hamming codes as linear codes. But before doing so, we introduce a few concepts from linear algebra that will make our life much easier when constructing the Hamming codes and analyzing their properties.

Linear Algebra on $\mathbb{Z}_2^n$

We first show how matrix multiplication can work in the type of linear space we consider here. It is almost the same as in the real-valued case introduced in appendix B.7, except for that here, everything is computed modulo 2. In short, for a $p \times q$ matrix A and a $q \times r$ matrix B, the product $A \cdot B$ is a $p \times r$ matrix, whose (i, j)th entry is defined as

$$(A \cdot B)_{ij} := \sum_{k=1}^{q} A_{ik} B_{kj} \bmod 2.$$

Let us consider an example. To multiply the 1×7 matrix (or vector)

$$x := \begin{pmatrix} 1 & 0 & 1 & 0 & 0 & 1 & 1 \end{pmatrix}$$

with the 7×3 matrix

$$H := \begin{pmatrix} 0 & 0 & 1 \\ 0 & 1 & 0 \\ 0 & 1 & 1 \\ 1 & 0 & 0 \\ 1 & 0 & 1 \\ 1 & 1 & 0 \\ 1 & 1 & 1 \end{pmatrix} \tag{C.1}$$

it is easiest to consider one column of H at a time. We have

$$x \cdot H = \begin{pmatrix} 0 & 1 & 1 \end{pmatrix}.$$

Here, we computed the first element as

$$1 \cdot 0 + 0 \cdot 0 + 1 \cdot 0 + 0 \cdot 1 + 0 \cdot 1 + 1 \cdot 1 + 1 \cdot 1 = 2 = 0 \bmod 2,$$

the second element as

$$1 \cdot 0 + 0 \cdot 1 + 1 \cdot 1 + 0 \cdot 0 + 0 \cdot 0 + 1 \cdot 1 + 1 \cdot 1 = 3 = 1 \bmod 2,$$

and the third element as

$$1 \cdot 1 + 0 \cdot 0 + 1 \cdot 1 + 0 \cdot 0 + 0 \cdot 1 + 1 \cdot 0 + 1 \cdot 1 = 3 = 1 \bmod 2.$$

The matrix H thus defines a map from $\mathbb{Z}_2^7$ to $\mathbb{Z}_2^3$, via $x \mapsto x \cdot H$. It is usually referred to as a *linear map*, because

$$(x + y) \cdot H = x \cdot H + y \cdot H.$$

We need two more concepts: linear independence of vectors and the dimension of a linear space. We call a set of vectors $x_1, \ldots, x_p \in \mathbb{Z}_2^n$ *linearly independent* if for $\lambda_1, \ldots, \lambda_p \in \mathbb{Z}_2$, we have

$$\lambda_1 x_1 + \lambda_2 x_2 + \cdots + \lambda_p x_p = 0 \quad \Longrightarrow \quad \lambda_1 = \lambda_2 = \cdots = \lambda_p = 0,$$

where, again, all computations are performed modulo 2. In words, there is only one way of combining the vectors $x_1, \ldots, x_p$ to get 0: We have to multiply each vector by 0. As an example, the vectors

$$x_1 = (1001),\ x_2 = (0101),\ x_3 = (1011),\ x_4 = (0010)$$

are not linearly independent (we then call them *linearly dependent*), since

$$1 \cdot x_1 + 0 \cdot x_2 + 1 \cdot x_3 + 1 \cdot x_4 = 0.$$

Given a a subset W of $\mathbb{Z}_2^n$ (e.g., our code), W is a *linear subspace* if the all-zero sequence is in W and

$$x, y \in W \quad \Longrightarrow \quad x + y \in W.$$

We can now define the *dimension* $\dim W$ of W. It is the maximum number of linearly independent vectors.[1] For example, the dimension of $\mathbb{Z}_2^4$ itself equals $\dim \mathbb{Z}_2^4 = 4$. This is the case,

1. Strictly speaking, one has to argue that this is a well-defined number by showing that any maximal set of linearly independent vectors has the same size. We are not going to prove that here, but we hope that you believe that this is the case.

because the unit vectors

$$e_1 := (0001),\ e_2 := (0010),\ e_3 := (0100),\ e_4 := (1000)$$

are linearly independent, and adding any other vector element would make the set linearly dependent (you may want to prove why this is the case). These maximal sets of linearly independent vectors are very practical. For example, one can show that each vector in the linear space can be written as a unique linear combination of such unit vectors (can you prove that, too?). For example, in $\mathbb{Z}_2^4$, the vector (1101) has the decomposition

$$1 \cdot e_1 + 1 \cdot e_2 + 0 \cdot e_3 + 1 \cdot e_4.$$

Among other things, this allows us to count the number of elements in W. Each factor can be either 0 or 1 and thus

$$\#W := 2^{\dim W}, \tag{C.2}$$

for example, $\#\mathbb{Z}_2^4 = 16$, which we know already, of course.

Construction of General Hamming Codes

Equipped with these tools, we can now define the Hamming code for $n = 2^m - 1$ using the $n \times m$ matrix

$$H := \begin{pmatrix} 0 & \cdots & 0 & 0 & 0 & 1 \\ 0 & \cdots & 0 & 0 & 1 & 0 \\ 0 & \cdots & 0 & 0 & 1 & 1 \\ 0 & \cdots & 0 & 1 & 0 & 0 \\ 0 & \cdots & 0 & 1 & 0 & 1 \\ 0 & \cdots & 0 & 1 & 1 & 0 \\ 0 & \cdots & 0 & 1 & 1 & 1 \\ 0 & \cdots & 1 & 0 & 0 & 0 \\ \vdots & \vdots & \vdots & \vdots & \vdots & \vdots \\ 1 & \cdots & 1 & 1 & 1 & 0 \\ 1 & \cdots & 1 & 1 & 1 & 1 \end{pmatrix}.$$

It is constructed by concatenating all $n = 2^m - 1$ binary words of length m as rows, while excluding the all-zero sequence. We define the code W as

$$W := \{x \in \mathbb{Z}_2^n : x \cdot H = 0\}.$$

Here, 0 stands for the all-zero sequence $(0 \cdots 0)$ that consists of n 0s; we can use this notation, since it is always clear from the context whether we mean the number $0 \in \mathbb{R}$ or the vector $0 \in \mathbb{Z}_2^n$. Equation (C.1) shows H for $n=7$ and $m=3$. In fact, the corresponding W is the Hamming code that forms the basis for the one we have shown in the figure on page 16. There we did not display the original codewords, but added the constant word (0110100) to avoid codewords like (0000000). Codeword 2 in the figure, for example, equals (0111011). And, indeed, for $x_2 = (0111011) - (0110100) = (0001111)$, we have $x_2 \cdot H = (0,0,0)$. How do we know that W is indeed a Hamming code? To prove this, we will exploit the fact that X is a linear code: $(x+y) \cdot H = x \cdot H + y \cdot H = 0$. This should not come as a surprise, given that we have gone through all this work!

Let us first prove that W is 1-error correcting. To do so, we must show that the Hamming distance between two codewords is at least 3:

$$\min_{x,y \in W} d(x,y) \geq 3.$$

Because W is linear and $d(0, y+x) = d(x+x, y+x) = d(x,y)$, this is equivalent to

$$\min_{y \in W} d(0,y) \geq 3.$$

Furthermore, $\min_{y \in W} d(0,y)$ equals the minimal number of linearly dependent rows in H. (This is shown by the following argument. Assume that the minimum is achieved for $y' \in W$. Then, y' has $d(0,y')$ nonzero entries, and since $y'H = 0$, the corresponding rows in H are linearly dependent. Conversely, a minimal set of linearly dependent rows can be represented by the ones in a vector $y' \in \mathbb{Z}_2$ that satisfies $y'H = 0$.) But clearly, the minimal number of linearly dependent rows of H equals 3: any

2 rows are linearly independent (no 2 rows are identical), and the first 3 rows are linearly dependent.

Finally, to prove that the code is perfect, we can again use a counting argument. We will show that each sequence has at most distance 1 to any of the codewords (that is, the balls with radius 1 span the whole space):

$$\bigcup_{x \in W} B_1(x) = \mathbb{Z}_2^n.$$

Note that we have[2] $\dim W = n - m$. We then have, using equation (C.2),

$$\# \bigcup_{x \in W} B_1(x) = \#W \cdot (n+1) = 2^{\dim W} \cdot 2^m = 2^{n-m} \cdot 2^m = 2^n = \#\mathbb{Z}_2^n,$$

which proves that the spaces must be equal.

The Hamming code W contains 2^{n-m} codewords. If the colors of the hats are distributed randomly, the probability of losing is, as stated in section 1.5,

$$P(\text{group loses}) = \frac{2^{n-m}}{2^n} = \frac{1}{2^m} = \frac{1}{n+1}.$$

The more players, the merrier!

C.2 CHAPTER 4: ANIMAL STICKERS AND CYCLIC GROUPS

Variation: Colored Hats with Infinitely Many Players

It is a mathematical curiosity that the colored-hats game still works with a countably infinite number of players and a finite number m of colors. This is, of course, irrelevant for all practical purposes, but we encourage you to indulge yourself with

2. The easiest way to see this is to use the rank-nullity theorem. It states that $\dim \mathbb{Z}_2^n = \dim \operatorname{im} H^\top + \dim \ker H^\top$, where $H^\top$ is the transpose of H, $\ker H^\top = W$ the kernel, and $\operatorname{im} H^\top$ is the image of $H^\top$. The dimension of $\operatorname{im} H^\top$ equals m, since H contains the linearly independent unit vectors e_i, $i = 1, \ldots, m$, as rows; that is, $H^\top$ has full column rank, in this case, m.

this thought experiment. It allows us to take a glimpse at the astonishing nature of the concept of infinity and the difficulties it raises.

Let us number the players with an increasing sequence, starting from 1 (the player in the back). The distribution of colors can then be represented by a sequence $(C_1, C_2, \ldots) = (C_k)_{k\in\mathbb{N}}$ of numbers $C_k \in \{0, \ldots, p-1\}$. It is now possible to construct an equivalence relation on the set of all such sequences:

$$(C_k)_{k\in\mathbb{N}} \sim (D_k)_{k\in\mathbb{N}} :\Leftrightarrow \text{only finitely many numbers are different.}$$

You can certainly convince yourself easily that this indeed defines an equivalence relation (which, in fact, is quite famous). Earlier we wrote down equivalence classes using so-called representatives. In $\mathbb{Z}/3\mathbb{Z}$, we wrote

$$\{\ldots, -1, 2, 5, 8, 11, 14, \ldots\} = [2],$$

for example; that is, we chose 2 as a representative of the corresponding equivalence class. Now, we will do the same for the infinite sequences and represent the (infinitely many) equivalence classes by

$$(C_k^1)_{k\in\mathbb{N}}, (C_k^2)_{k\in\mathbb{N}}, (C_k^3)_{k\in\mathbb{N}}, \ldots.$$

This seems like a natural and harmless step, but it is far from being that. To guarantee that such a choice is possible requires the so-called "axiom of choice," which has been the topic of many debates and mathematical paradoxes. When encountered for the first time, it can surprise you by showing how much mathematics is beyond the classes of "correct" and "incorrect."

For now, let us assume that we do have access to the representatives $(C_k^1)_{k\in\mathbb{N}}, (C_k^2)_{k\in\mathbb{N}}, (C_k^3)_{k\in\mathbb{N}}, \ldots$. Suppose that the current game is described by the true sequence $(C_k)_{k\in\mathbb{N}}$. The key observation for solving the (imaginary) game is that each player does not know the true sequence $(C_k)_{k\in\mathbb{N}}$ but recognizes its equivalence class,

$$[(C_k)_{k\in\mathbb{N}}]=[(C_k^{480213})_{k\in\mathbb{N}}],$$

say. Since $(C_k)_{k\in\mathbb{N}}$ and $(C_k^{480213})_{k\in\mathbb{N}}$ differ by only finitely many numbers, there is a number M, say, beyond which they are identical. The key idea is that any player a for whom the remaining sequence of hat colors $(C_k)_{k\geq a+1}$ they see matches the sequence of the representative $(C_k^{480213})_{k\geq a+1}$ of the equivalence class, guesses his own color to be the one from the representative (i.e., C_a^{480213}). Therefore, players $M+1, M+2, \ldots$ will announce the correct color. What about the first M players? Since the number M is known to players $1, \ldots, M-1$, we can apply the solution from the finite case with $M-1$ players. Again, the first player computes the corresponding sums of the observed colors from $2, 3, \ldots, M-1$, ignoring everything that comes after $M-1$. This way, players $2, 3, \ldots, M-1$ receive sufficient information to announce their hat colors correctly, as do the players $M+1, M+2, \ldots$. Player M's answer is certainly incorrect, and thus, the expected number of false answers is $2-1/m$. If this thought experiment feels a bit unsettling, we encourage you to take a walk outside. The world is beautifully finite.

C.3 CHAPTER 5: OPERA SINGERS AND INFORMATION THEORY

Uniform Distributions and Entropy

We will now prove that the uniform distribution indeed maximizes the entropy. Consider a random variable that takes the values $x_1, \ldots, x_m$ with probabilities $p_1, \ldots, p_m$, respectively. We now want to prove that $H(p_1, \ldots, p_m)$ is maximized for the uniform distribution (that is, if all of the p_i equal $1/m$).

The statement follows from a well-known mathematical inequality, called the *Gibb's inequality*. We start with the observation that for all $x>0$, we have

$$\log_2 x \leq \frac{x-1}{\ln 2} \tag{C.3}$$

with equality if and only if $x=1$. Here, we write $\ln 2 := \log_e 2$ to stress that the logarithm is taken to the base $e=\exp(1)$. There are several ways of seeing why equation (C.3) holds. It is probably easiest if you draw the two graphs. Regarding a formal proof, if you know what "strong convexity" means, you can also prove that $x \mapsto \ln x$ is strictly convex with $x \mapsto x-1$ being the tangent at the graph at $x=1$. From equation (C.3), it follows that, for any i, we have

$$p_i \log_2 \frac{1}{mp_i} \leq p_i \frac{\frac{1}{mp_i}-1}{\ln 2}$$

with equality if and only if $mp_i=1$. This even holds if we sum over all i. That is,

$$\sum_{i=1}^{m} p_i \log_2 \frac{1}{mp_i} \leq \sum_{i=1}^{m} p_i \frac{\frac{1}{mp_i}-1}{\ln 2}$$

with equality if and only if for all i, $p_i = 1/m$. Transforming both sides yields

$$-\log_2 m + \sum_{i=1}^{m} p_i \log_2 \frac{1}{p_i} \leq \frac{1-1}{\ln 2} = 0$$

with equality if and only if for all i, $p_i = 1/m$. This proves that

$$H(p_1, \ldots, p_m) \leq \log_2 m$$

with equality if and only if for all i, $p_i = 1/m$. Thus, the entropy is maximized for the uniform distribution.

Stepwise Strategies Are Not Always Optimal

In section 5.4, we proposed to optimize the average or minimal information content step-wise, which is often called a "greedy" strategy. We now provide an example that shows that such an approach does not necessarily yield an optimal strategy. Consider the set of hypotheses

$$\Omega = \{1, 2, 3, \ldots, 24\}.$$

Suppose that there are only 4 questions that we can ask:

$$O_1 : \left\{ \begin{array}{rrrrrrrrrcl} 1, & 2, & 3, & 4, & 5, & 6, & 7, & 8, & 9 & \mapsto & 0 \\ 10, & 11, & 12, & 13, & 14, & 15, & 16, & 17, & 18 & \mapsto & 1 \\ & & & 19, & 20, & 21, & 22, & 23, & 24 & \mapsto & 2 \end{array} \right.$$

$$O_2 : \left\{ \begin{array}{rrrrrrrrrcl} & & 1, & 2, & 3, & 10, & 11, & 12, & 21 & \mapsto & 0 \\ 4, & 5, & 6, & 13, & 14, & 15, & 22, & 23, & 24 & \mapsto & 1 \\ & 7, & 8, & 9, & 16, & 17, & 18, & 19, & 20 & \mapsto & 2 \end{array} \right.$$

$$O_3 : \left\{ \begin{array}{rrrrrrrrrcl} 1, & 4, & 7, & 10, & 13, & 16, & 19, & 21, & 22 & \mapsto & 0 \\ & 2, & 5, & 8, & 11, & 14, & 17, & 20, & 23 & \mapsto & 1 \\ & & 3, & 6, & 9, & 12, & 15, & 18, & 24 & \mapsto & 2 \end{array} \right.$$

$$O_4 : \left\{ \begin{array}{rrrrrrrrrcl} & 1, & 4, & 6, & 7, & 8, & 9, & 12, & 16 & \mapsto & 0 \\ & 5, & 11, & 14, & 15, & 17, & 18, & 19, & 20 & \mapsto & 1 \\ & 2, & 3, & 10, & 13, & 21, & 22, & 23, & 24 & \mapsto & 2 \end{array} \right.$$

We have $\#\Omega = 24$, and the questions have 3 possible answers. We thus need at least 3 questions to find the secret number ω^*.

The strategy maximizing both the average and the minimal information content in each step would clearly tell us to start with the question O_4. All of its outcomes, $P(O_4 = 0) = P(O_4 = 1) = P(O_4 = 2) = 8/24 = 1/3$, have equal probability, and the average (or minimal) information equals $\log_2 3 \approx 1.585$ bits. Starting with O_4, however, will not allow us to always identify the secret ω^*. Suppose that the answer to O_4 equals $O_4(\omega^*) = 2$. If we then continue with O_1, and we receive the answer $O_1(\omega^*) = 2$, no matter whether we continue with O_2 or O_3 as a third question, there is a pair of two different ω values that we cannot distinguish between. This is shown by the first 4 rows of the following table:

ω^*	O_4	O_1	O_2	O_3
23	2	2	1	✗
24	2	2	1	✗
21	2	2	✗	0
22	2	2	✗	0
13	2	✗	1	0
22	2	✗	1	0

In fact, the table shows that in our example, there is no strategy starting with O_4 that is certain to identify ω^*. Even if we continue with O_2 rather than O_1 (and then use O_3 or O_1) or if we continue with O_3 (and then use O_1 or O_2), there are always two different ω^* values that yield the same answers.

If we start with O_1 instead of O_4, however, and use the questions O_1, O_2, O_3, we are guaranteed to always identify ω^*: There are no two ω^* values that map to the same 3-tuple.

This counterexample is constructed to show that we cannot expect optimality when performing a greedy search strategy. In most practical situations, however, the greedy approach might be a good way to start.

C.4 CHAPTER 6: ANIMAL MATCHING AND PROJECTIVE GEOMETRY

In the incidence matrix A, each row corresponds to a player, and each column corresponds to an animal. A value of 1 in position (k, j) of the incidence matrix (we write $A_{k,j} = 1$ in this case) indicates that player k has taken animal j.

In the remainder of this section, we want to relate the properties of A to the constraint in the game that every pair of players needs to hold at least 1 animal in common. We also make an argument for why all 7 players have to hold exactly 3 animals for a valid solution.

First, $A^\top$ is the so-called "transpose" of A. The transpose $A^\top$ can be thought of as A with interchanged rows and columns.

The entry (k, j) in matrix $A^\top$ is equal to the entry (j, k) in A. The dimensions of A and $A^\top$ are only the same if the number of rows and columns is equal (which is true here). An interesting matrix to study is

$$AA^\top,$$

the matrix multiplication of A with $A^\top$ (see appendix B.7 for a definition of a matrix multiplication). The matrix $AA^\top$ has the entries (using the definition of a transpose)

$$(AA^\top)_{k,j} = \sum_{\text{animal } \ell} A_{k,\ell}(A^\top)_{\ell,j} = \sum_{\ell} A_{k,\ell}A_{j,\ell}.$$

The entry (k, j) of matrix $AA^\top$ thus uses rows k and j of A, corresponding to players k and j, and is summing over all columns (animals) the product $A_{k,\ell}A_{j,\ell}$ of the corresponding entries of A. This product is usually 0 and can only be 1 if both $A_{k,\ell} = 1$ and $A_{j,\ell} = 1$ (that is, both players need to have animal ℓ in their possession). By summing over all animals, we get the total number of animals that both players have in common. If we want to guarantee that every pair of players has at least 1 animal in common, then we need to make sure that for all $k, \ell \in \{1, \ldots, 7\}$,

$$(AA^\top)_{k,\ell} \geq 1.$$

The diagonal part of $AA^\top$ will always be equal to the diagonal part of $3 \cdot \text{Id}$ (that is, a matrix with 0s everywhere except for entries equal to 3 on the diagonal), as each row of A contains exactly three 1s. Assume that the incidence matrix satisfies

$$AA^\top = 2 \cdot \text{Id} + J,$$

where Id is the identity matrix (1s on the diagonal and 0s everywhere else), and J is a matrix with all entries identically equal to 1. Then all pairs of players have 1 animal in common. Then the way that the audience picks the pairs does not matter, and the players will always win.

Is there another possible solution where, for example, some entries in $AA^\top$ take the value 2, so that some pairs of players have 2 animals in common? To answer the question, it helps to look at the sum over all entries of $AA^\top$. In any arrangement where each player has 3 distinct animals, each of the 7 animals is present for 3 players. There will thus always be $3^2 = 9$ player-pairs (k, ℓ) that share the same animal, counting also cases where $k = \ell$ (diagonal entries of $AA^\top$) and counting both (k, ℓ) and (ℓ, k) if $k \neq \ell$ (that is, looking at upper-triangular and lower-triangular parts of $AA^\top$). Each animal thus contributes $3^2 = 9$ to the sum of all entries of $AA^\top$. Summing over all animals, we always get the value $7 \cdot 3^2 = 63$ for the sum of $AA^\top$ over all entries. This sum-constraint is satisfied, for example, by the solution $AA^\top = 2 \cdot \mathrm{Id} + J$. If we now find a solution that deviates from $AA^\top = 2 \cdot \mathrm{Id} + J$ by, for example, allowing some pairs to have 2 or more animals in common (which implies a value of 2 or more on the off-diagonal elements of $AA^\top$), then the sum constraint implies that this value has to be compensated for by some pairs not having any animal in common, with a corresponding entry of 0 on the off-diagonal elements. Hence $AA^\top = 2 \cdot \mathrm{Id} + J$ needs to be fulfilled for any solution that guarantees success.

The task is thus to create an incidence matrix A of dimension 7×7 that satisfies $AA^\top = 2 \cdot \mathrm{Id} + J$. In other words, the total number of animal matches is just enough to give each pair of players 1 animal to match. If any pair of players has more than 1 animal in common, then other pairs will have no animal in common.

C.5 CHAPTER 8: THE FALLEN PICTURE AND ALGEBRAIC TOPOLOGY

We now provide a formal introduction to the fundamental group. This concept is certainly not necessary for understanding chapter 8, but if you have some mathematical training, you might be wondering how to formalize the concepts introduced in chapter 8. In particular, if you know when functions $f : [0, 1] \to \mathbb{R}^n$ and $F : [0, 1] \times [0, 1] \to \mathbb{R}^n$ for some $n \in \mathbb{N} \setminus \{0\}$ are said to be continuous, we hope that the following sequence of definitions will make sense.

Let X be a space that is a subset of $\mathbb{R}^n$ for some $n \in \mathbb{N} \setminus \{0\}$. We have the following definitions. A *loop around* $x_0 \in X$ (for simplicity, we will simply call this a "loop") is a continuous map

$$f : [0,1] \to X$$

with $f(0) = f(1) = x_0$. A *homotopy of loops* is a collection of loops

$$f_t : [0,1] \to X, \quad 0 \leq t \leq 1,$$

such that the map

$$F : \begin{array}{ccc} [0,1] \times [0,1] & \to & X \\ (s,t) & \mapsto & f_t(s) \end{array}$$

is continuous. Two loops g and h are *homotopic* if there is a homotopy $(f_t)_t$ with

$$f_0 = g \qquad \text{and} \qquad f_1 = h.$$

This defines an equivalence relation $\sim$ between loops. We identify loops with their equivalence classes. The concatenation of loops defines a group action:

$$g \circ h : \begin{array}{ccc} [0,1] & \to & X \\ t & \mapsto & \begin{cases} g(2t) & \text{if } t \leq 0.5 \\ h(2t-1) & \text{otherwise.} \end{cases} \end{array}$$

The *fundamental group* $\Pi_1(X, x_0)$ is defined as the set of all loops modulo the equivalence relation $\sim$ with concatenation as the group action. One can show that for path-connected spaces X, the fundamental group does not depend on x_0.

REFERENCES

Aldous, D. Random walks on finite groups and rapidly mixing Markov chains. *Séminaire de Probabilit XVII*, pages 243–297, 1983. Lecture Notes in Mathematics 986. Springer, Berlin, Heidelberg.

Aldous, D., and P. Diaconis. Shuffling cards and stopping times. *American Mathematics Monthly*, 93:333–348, 1986.

Aravamuthan, S., and S. Lodha. Covering codes for hats-on-a-line. *Electronic Journal of Combinatorics*, 13(1):21, 2006.

Aspnes, J., R. Beigel, M. Furst, and S. Rudich. The expressive power of voting polynomials. *Combinatorica*, 14(2):135–148, 1994.

Baldwin, R. R., W. E. Cantey, H. Maisel, and J. P. McDermott. The optimum strategy in blackjack. *Journal of the American Statistical Association*, 51:419–439, 1956.

Bayer, D., and P. Diaconis. Trailing the dovetail shuffle to its lair. *Annals of Applied Probability*, 2:294–313, 1992.

Bruck, R. H., and H. J. Ryser. The nonexistence of certain finite projective planes. *Canadian Journal of Mathematics*, 1:88–93, 1949.

Buhler, J. P. Hat tricks. *Mathematical Intelligencer*, 24(4):44–49, 2002.

Candès, E. J., J. Romberg, and T. Tao. Robust uncertainty principles: Exact signal reconstruction from highly incomplete frequency information. *IEEE Transactions on Information Theory*, 52(2):489–509, 2006.

Cover, T. M., and J. A. Thomas. *Elements of Information Theory*. Wiley-Interscience, New York, NY, 2006.

Curtin, E., and M. Warshauer. The locker puzzle. *Mathematical Intelligencer*, 28(1):28–31, 2006.

Demaine, E. D., M. L. Demaine, Y. N. Minsky, J. S. B. Mitchell, R. L. Rivest, and M. Patrascu. Picture-hanging puzzles. ArXiv e-prints (1203.3602), 2012.

Diaconis, P. *Group Representations in Probability and Statistics*. Lecture Notes—Monograph Series, Institute of Mathematical Statistics, Hayward, CA, 1988.

Donoho, D. L. Compressed sensing. *IEEE Transactions on Information Theory*, 52(4):1289–1306, 2006.

Ebert, T. Applications of recursive operators to randomness and complexity. PhD thesis, University of California, Santa Barbara, 1998.

Euler, L. *Novi commentarii academiae scientiarum petropolitanae*. Reprinted in *Opera Omnia*, serie prima, vol. 26 (A. Speiser, ed.), n. 325, 139–157, 1776.

Eves, D. Problem E712: The extended coin problem. *American Mathematical Monthly*, 53:156, 1946.

Fischer, G. *Lineare Algebra*. Vieweg, Braunschweig/Wiesbaden, Germany, 13 ed., 2002.

Gál, A., and P. B. Miltersen. The cell probe complexity of succinct data structures. In *International Colloquium on Automata, Languages, and Programming*, pages 332–344. Springer, New York, 2003.

Gilbert, E. Theory of shuffling, 1955. Technical memorandum, Bell Labs, Murray Hill, NJ.

Goyal, N., and M. Saks. A parallel search game. *Random Structures & Algorithms*, 27(2):227–234, 2005.

Hardin, C., and A. Taylor. An introduction to infinite hat problems. *Mathematical Intelligencer*, 30:20–25, 2008.

Hatcher, A. *Algebraic Topology*. Cambridge University Press, Cambridge, UK, 2002.

Hughes, D. R., and E. C. Piper. *Design Theory*. Cambridge University Press, Cambridge, UK, 1985.

Kåhrström, J. On projective planes. C-Uppsatts Mid Sweden University; website (accessed 17.09.2019), 2002. http://kahrstrom.com/mathematics/documents/OnProjectivePlanes.pdf.

Lam, C. W. H. The search for a finite projective plane of order 10. *American Mathematical Monthly*, 98(4):305–318, 1991.

MacKay, D. J. C. *Information Theory, Inference & Learning Algorithms*. Cambridge University Press, New York, 2002.

Paterson, M. B., and D. R. Stinson. Yet another hat game. *Electronic Journal of Combinatorics*, 17(1):R86, 2010.

Pegg Jr., E. Hanging picture. Website (accessed 31.07.2019), 2002. http://www.mathpuzzle.com/hangingpicture.htm.

Polster, B. The intersection game. *Math Horizons*, 22(4):8–11, 2015.

Schell, E. D. Problem E651—weighed and found wanting. *American Mathematical Monthly*, 52:42, 1945.

Seifert, H. Konstruktion dreidimensionaler geschlossener Räume. PhD thesis, Technische Hochschule Dresden, Dresden, 1931. JFM 57.0723.01 Berichte Leipzig 83, 26–66.

Shannon, C. E. A mathematical theory of communication. *Bell System Technical Journal*, 27(3):379–423, 1948.

Stark, D., A. Ganesh, and N. O'Connell. Information loss in riffle shuffling. *Combinatorics, Probability and Computing*, 11(1):79–95, 2002.

Trefethen, L. N., and L. M. Trefethen. How many shuffles to randomize a deck of cards? *Proceedings: Mathematical, Physical and Engineering Sciences*, 456(2002):2561–2568, 2000.

Van Kampen, E. R. On the connection between the fundamental groups of some related spaces. *American Journal of Mathematics*, 55(1):261–267, 1933.

INDEX

Below, you can find the page(s) on which the listed terms appear. Boldface numbers refer to the page where terms are defined.